MUSIC FROM THE ROAD

MUSIC
FROM THE
ROAD

Views and Reviews 1978–1992

TIM PAGE

New York Oxford
OXFORD UNIVERSITY PRESS
1992

Oxford University Press

Oxford New York Toronto
Delhi Bombay Calcutta Madras Karachi
Kuala Lumpur Singapore Hong Kong Tokyo
Nairobi Dar es Salaam Cape Town
Melbourne Auckland

and associated companies in
Berlin Ibadan

Copyright © 1992 by Tim Page

Published by Oxford University Press, Inc.,
200 Madison Avenue, New York, New York 10016

Oxford is a registered trademark of Oxford University Press

Library of Congress Cataloging-in-Publication Data
Page, Tim.
Music from the road : views and reviews, 1978–1992
Tim Page.
p. cm. Includes index.
ISBN 0-19-507315-0
1. Music—20th century—History and criticism.
2. Musicians—Interviews.
3. Music—Reviews.
I. Title. ML60.P142 1992
780′.9′04—dc20 91–24134

9 8 7 6 5 4 3 2 1
Printed in the United States of America
on acid-free paper

For Vanessa

CONTENTS

INTRODUCTION

One evening in 1979, a few weeks out of college and thoroughly confused about the future, I walked into a Greenwich Village store and purchased Pierre Boulez's new recording of the complete music of Anton Webern. I took it home, listened to it, loved it, and then spent three days putting my reactions into words. When the story was finished, I sent it, unsolicited, to the *Soho News,* a lively, hedonistic, arts-oriented weekly newspaper published in lower Manhattan.

I had no connections at the *Soho News,* no reason to believe that the editors would want my thoughts on this particular subject (then, as now, dissertations on Webern were not exactly hot property). But they accepted the story, printed it, even paid me something eventually. And suddenly I was a music critic.

One can make generalizations about the practitioners of many professions, but each music critic has a distinct history. Some attended conservatories, others did not. Some began as musicians and gravitated to writing; others were writers who decided to specialize in music. Rightly or wrongly, there is no qualifying examination, no "bar" to pass.

Still, as Virgil Thomson pointed out half a century ago, there are two prerequisites: It is necessary to know about music and to know how to write. Fine musicians often produce dreary prose, while few professional writers have the musical experience that would enable them to function as music critics. Which of the two skills is the more important? A musician would likely rank technical knowledge as paramount, while an editor would be inclined to go with the best possible writer for a position as critic.

It is rare to find one person in whom both skills are honed to an equal degree: Usually, a critic is either a "better" writer than listener or a "better" listener than writer. Olin Downes, who covered music for the *New York Times* from 1924 to 1955, is probably the classic example of a critic who heard better than he wrote. His opinions generally hold up (particularly his estimations of Mahler and Sibelius) but his language will likely impress a contemporary reader as turgid, clotted, all but impenetrable. B.H. Haggin, on the other hand, sometimes seems to have been "wrong" about a lot of things, but the clean, cutting exactitude of his early writing is often masterly and instructive.

A critic's work will naturally reflect his schooling. My principal interest (and training) is in the music of our time, in new work, structures, sonic innovation. Indeed, it was after the world premiere of Steve Reich's "Music for 18 Musicians" in April 1976 that I wrote my first more-or-less mature criticism. I returned home, exhilarated by what I had heard, with an urgent need to *react*, in some concrete and personal manner, to Reich's work. If I had been a dancer, I might have danced it; if I had been a filmmaker, I might have plotted a film to it. As a would-be writer, I had to try and write it.

So I went to the typewriter and started to free associate. Entranced, I typed page upon page of copy, gradually zeroing in on the images that came to mind. What I had heard struck me as so unusual, so beautiful, so distinctly of its time but far from typical "new music," that I wrote through the night, trying to summarize my impressions. Five years later, I published a study of "minimalist" music in *High Fidelity* (which I now consider seriously flawed and have not included in this collection). It would incorporate some of what I wrote that night:

> Minerva-like, the music springs to life fully formed—from dead silence to fever pitch. There is a strong feeling of ritual, a sense that on some subliminal plane the music has always been playing and that it will continue playing forever . . . Imagine concentrating on a challenging modern painting that becomes just a little different every time you shift your attention from one detail to another. Or trying to impose a frame on a running river—making it a finite, enclosed work of art yet leaving its kinetic quality unsullied, leaving it flowly freely on all sides. It has been done. Steve Reich has framed the river.

(My writing has not changed dramatically since then and I am well aware of its strengths and weaknesses. But I also know now that, for better and for worse, this is the way I write—my "style," if you like. I can hope that it will continue to evolve but I cannot radically alter it without cutting it off at the source. If my writing is tremulous and theatrical at times, I hope there are other occasions when it is colorful and alive.)

In 1979, shortly after the *Soho News* accepted my Webern piece, I was appointed classical music editor for the munificent sum of $100 a week. I remained with the paper for the next two and a half years—indeed, for the rest of its life—and it was a perfect place to make those vital, nourishing mistakes through which one learns. There one found the voices of neo-conservatives and Marxists, radical feminists and libertines, chronic potheads and anti-drug crusaders—the common bonds being only a fierce, eccentric independence and an abiding love for the *Soho News*. At its best, the paper was full of internal tension and contrast, yet forged a strange synthesis from the chaos, with everybody's personal vision of Utopia treated with equal indifference.

But the early 80s were a disastrous time for print journalism. One afternoon in March 1982 the staff was summoned to a meeting and told that the *Soho News* had folded, that we should gather our belongings and leave the building. Floundering and broke, I was rescued by the *Saturday Review*, which hired me to replace Irving Kolodin as its music critic but the magazine ceased operations five months later. And then one day, as I was beginning to wonder how I was going to pay a looming American Express bill, John Rockwell asked me if I would come on as the regular stringer at the *New York Times*.

Thus began five extraordinary, frustrating and productive years. There is no place like the *Times*—I say that with awe, gratitude, respect and a certain relief that I am no longer there. Suddenly my life was a whirl of concerts, deadlines, press releases, long-lost friends from music school who suddenly reappeared, hostile letters from prim doctoral students on whose pet project I had stepped, queries from Taiwan about where to find a record in Taipei and an incredible, unrelenting pressure. Nobody cared very much if the *Soho News* liked or didn't like a concert, but now the words that I wrote were scrutinized and debated in Lincoln Center restaurants and 57th Street management offices, praised and denounced by interested parties and the musically literate throughout the world. There is a coterie of readers who take the *Times* very seriously: I recall a lengthy missive from an English professor who claimed I had misused the comma. Make a genuine factual *error* and the sky falls.

It is a first-class seasoning. I will admit that I was not prepared to take the job when it came my way; nothing can really prepare a writer for work on the *Times*. One steps into the maelstrom, then sinks or swims. Slowly, I learned to swim.

I had the naive idea that a *Times* position would be glamorous and exciting. Exciting, yes, but I do not recall the glamour so much as the intensity, the continual need to produce, the sheer hard work of it all. Once, in the middle of a thirty-minute flight from Atlanta to Charleston, I realized that I had neglected to write a review of the concert I'd heard the night before. I asked the flight attendant for some paper but the skies grew choppy and we were all confined to our seats for the duration of the trip. I knew that my editor would expect me to phone in my review the moment we touched down and that I would likely commit the cardinal sin of journalism—missing a deadline—unless I set to work immediately. So I rummaged through the seat pocket in front of me; in desperation, I fetched out a motion sickness bag. And there I wrote the review which ran in the next morning's *Times*.

This anecdote encapsulates the combination of deadline tensions and chaotic circumstances from which newspaper writers must synthesize a paragraph or two that can satisfy editor and reader (and, less often, themselves and their subjects as well). There is nothing of the ivory tower to daily journalism. Writing hastily about a performance or a work of art is a terrifying responsibility, but it is also a craft in which one grows more assured as one practices it.

There were weekends when I would cover as many as eight individual concerts; my record for a week's work was thirteen. (The *Times* has since radically scaled back its music coverage, to the point where all the critics combined sometimes write fewer than thirteen.) I quickly learned to take copious notes; otherwise the details would be blurred by the time I sat down to write. I forgot almost everything once it was over anyway—in a sort of purging process—and today I remember only the very good and the very bad concerts that I heard. I regularly walk past posted advertisements and notice some blurb that says something like " . . . remarkable musicianship . . . Tim Page, *New York Times*" and draw a complete blank. On other occasions, I will meet somebody who wants to discuss an opinion I held years ago. And I will be incapable of saying anything authoritative; I can neither defend nor amplify my comments because I simply don't recall the circumstances under which they were made.

I *do* remember the sixtyish soprano who decided, at the last minute to dedicate her concert to the victims of the KAL Flight 007 shot down over Russia. (To emphasize her sorrow, she bused in a group of Korean nuns who wept dutifully as she warbled the Bach-Gounod "Ave Maria.") I remember the venerable pianist who had what can only be described as a complete nervous breakdown onstage at Alice Tully Hall. And, of course, I also remember the magnificent moments: Mieczyslaw Horszowski's recitals at Town Hall, played when the pianist was in his late nineties; Frederica von Stade's "Non so piu" at the Met; Karajan's visits with the Berlin Philharmonic, and many more.

Because I was the junior critic at the *Times*, I received most of the junior assignments. This meant a lot of marginal new music concerts (for which I was grateful—they were more engaging than most of the more traditional events) and a lot of debuts.

Covering debuts could be genuinely exciting. It was highly satisfying to leave a first-class performance by a young musician and write something that would tell the world about it. (Maria Bachmann, profiled in this book, was my favorite "discovery.") But, more often, debuts were somewhat frustrating because there was little to say, for good or for ill, about what had taken place. It was necessary to file a report that was (a) accurate (b) as humane as possible and (c) of at least vague interest to the disinterested reader.

I have always taken most pleasure in writing about 20th-century music, but any newspaper critic must be a generalist as well. At the *Soho News* I had made my own assignments; at the *Times* I was told where to go and I went. I might attend a concert of African drumming one night and the Metropolitan Opera the next. Once, I covered a concert of Mozart chamber works, a program of computer music and a tuba recital, all in the same long day. And then there were feature stories to write—about celebrated artists preparing their annual Carnegie appearance, about the Musicians' Union and its dictates, about the upcoming season and, of course, the occasional obituary.

A critic quickly learns that unfavorable reviews, especially of the glib, dismissive "Joe Jones played Mozart last night; Mozart lost" variety, are by far the easiest to write, but a string of insults is scarcely criticism. It is much more challenging to say something serious about a performance that has been moving and effective. Most difficult of all are the reviews of little events like, say, a Telemann chamber program in a church basement. It is a genuinely good thing that such concerts exist: they are unpretentious, they are usually reasonably well played and they bring happiness to people in the neighborhood. But there is nothing much to say about them that will not come across as damning with faint praise.

Two myths about the profession need debunking. It is true that few of us began with the determination to become critics but I have yet to find a single example of one durable stereotype—that of the failed artist turned destructive critic savaging anything talented in his way. (Musicians like to believe this sort of thing, as a way to justify negative reviews.) Nor have I found any of my colleagues to be personally dishonest. I have suspected them of having tin ears now and again, but I have never found evidence of corruption. And I have heard of no bribes changing hands in the years I've been on the beat.

Almost everybody in the music business is here because of a love— a calling, if you like. One must assume, until proven otherwise, that composers and performers are acting in good faith, with aspirations to artistic nobility. Because there is not the same possibility of a fast buck that we find in many of the other arts, the sort of barbed, slash-and-run criticism that we find in some movie, book, theater and commercial music reviews has no legitimate place in our discipline. It should be obvious that reviewing the latest trite film or pulp novel—that is, purely commercial material—is a different case from reviewing some scared young music student in a small auditorium.

This is not to advocate a bland, uncritical criticism. Journalists have a duty to tell the truth, and the exposure of incompetence goes along with the job. Still, the reviewer does not work in a vacuum. What he writes will have an effect, and the words should be chosen with the utmost care. It is all a question of degree—one can say anything one wants but the tone should be adapted to suit the circumstances. There is no reason to break the spirit of a serious young artist in print. If the concert has been less than satisfactory, there are ways of getting that across without mockery or condescension, without making a liar of one-self, but without being unnecessarily cruel.

Reading criticism of the past should convince us that there is no need to champion one single cause or aesthetic. The heated attempts to prove Wagner or Brahms the more important composer, or Toscanini or Furtwängler the better conductor seem futile in retrospect. The four, all of them great musicians, approached their craft in different ways. Likewise I do not believe that Elliott Carter and Philip Glass—radically dissimilar creators—need do aesthetic battle with one another. They have chosen divergent paths, each of which, in my opinion, has its own merit. As the proverb states, there are many paths to truth.

Since 1985 I've been teaching criticism, first at the Manhattan School of Music, later at the Juilliard School. My classes begin with an exercise that is meant to put a student directly into the role of a critic. I play a recording of a piece that is likely to be both unfamiliar and difficult to place—something like the second movement of the Sibelius 4th Symphony or the Interlude from Vaughan Williams' "Sinfonia Antartica." Then the class is asked to write a brief essay about what it has heard, allowing only ten minutes. After some groaning about the "impossibility" of the assignment, the class eventually comes through.

I collect the papers, then play the music again, this time naming it and identifying the composer. The students write another ten-minute essay. Finally, I give a detailed introduction to the piece, answer any questions, play the record again, and request yet another essay.

The results of this exercise are revealing, for the initial, untutored response to the music is almost always the most original and perceptive. The moment I tell the class that they have been listening to Sibelius, phrases like "the austere Finnish master" and platitudes about northern skies start creeping into the writing. And then, after the third exposure, the students essentially recycle the program notes I have fed them. They seem to be more secure in parroting than in venturing any opinions of their own.

The ideal, of course, is to balance an intimate knowledge of the score with an eternal capacity to be surprised. A good critic can listen to, say, the Sibelius Fourth, know it by heart, know all about its history, all about the opinions it has provoked, and still reflect anew and come to some fresh conclusions. But first one must learn to trust one's ears.

Ultimately, explication is the essence of a critic's job. Stating that we liked something or didn't is about one-tenth of the matter; explaining how that conclusion was reached is vastly more important.

Novice critics see the world in black and white, good and bad; only later do they begin to recognize shades of gray. Still further on, simple judgments of good and bad give way to the realization that there are at least four distinct polarities—masterpieces that one loves, masterpieces that one dislikes, flawed works that one loves and flawed works that one dislikes—and, of course, everything in-between. (I recognize, for example, that Debussy's "Pelleas et Melisande" is a unique and original work of art but it moves me not at all, while something so obviously flawed as Busoni's Piano Concerto "gets" me every time.)

I have brought in several colleagues to talk to the classes. The critical community is a very small group: We know and generally like one another but often strongly disagree. There is, to put it mildly, a great deal of contention about some performers and composers. Some artists attract nothing but passionate admirers and detractors (for example, there seem to be only two opinions about Philip Glass and Nadja Salerno-Sonnenberg). On the other hand, there seems to be very little controversy about mainstream artists such as Perlman and Galway, but not many critics would choose them as personal favorites. A certain deviation from the mean is prized.

There is sometimes little disagreement about the qualities of an artist, but a considerable diversity in the response to those qualities. Once I compared the *Times* reviews of a famous pianist. He had received many notices over the course of thirty-five years and they were of two minds. One school said that he played with superhuman strength, like a powerful machine, and that the results were magnificent. The other group held that he played with superhuman strength, like a powerful machine, and that the results were antimusical. Half the critics were thrilled, half offended, but the actual descriptions of the performances were not dissimilar.

Critics never discuss performances at intermissions: It is considered faintly suspicious and can be misunderstood as an attempt at coercion. More likely, it is a sign of insecurity—the fishing of a young critic for a consensus so that his review will not be too far off center and alert his editors to his perceived deficiencies. Beginners often feel the need to verify what they've heard. Later on we become proud of our passions, quirks and idiosyncrasies, and flaunt, instead of conceal, them. I doubt that there is anyone in the business with whom I could not get into a whale of an argument if I tried.

Since moving to *Newsday,* I have much enjoyed participating in the creation of a cultivated, rather less Jovian, "other" voice in New York. It has been a period of immense personal and professional satisfaction. My editors have allowed me maximum latitude in developing a department that would give the *Times* some real competition in daily music coverage, for the first time since the demise of the *Herald Tribune.* Indeed, I'm looking toward the day, not too long from now, when the *Times* will be the "other" voice in New York.

It has now been a dozen years since I published my first professional criticism, and the time seemed ripe for a summing up. And so I have read through my "compleat" works. It has been a sobering experience: Of the 100 or so pieces I wrote for the *Soho News,* the 1500-odd reviews and innumerable feature pieces written for the *Times,* the roughly 800 articles I have fashioned for *Newsday* and 200 or 300 freelance articles, I found very little that I wanted to save (or, for that matter, wanted anyone to *see,* ever again.) But this is always the fate of journalism— "Who wants yesterday's papers?" and all that.

Still, here are some articles and reviews that I hope will have some continuing appeal. Many of them are about contemporary music, which remains my principal interest. I believe that a classical critic must be seriously involved with the music of his time: The premiere of a new symphony is at least as important as the latest professional runthrough of Beethoven's Fifth.

It should be remembered that this book is not an attempt to provide a complete survey of the musical scene, but rather a collection of what one observer considers his livelier articles. I hope the best is yet to come.

New York City   Tim Page
March 3, 1992

1 MARIN ALSOP

Half past six in Eugene, Ore., and Marin Alsop—the leader of the city's symphony and, closer to home, the new music director of the Long Island Philharmonic—is meeting with conducting students from the University of Oregon.

There are a dozen of them, more or less—healthy, eager, fresh-faced and slightly damp from the Wilamette Valley's eternal drizzle. Significantly, half of them are women. Such a percentage would have been unheard of ten years ago, when Alsop was pounding on conservatory doors throughout the Northeast looking for somebody willing to teach her conducting, willing to give her that one chance....

Alsop, thirty-three, comes on like one of the gang, a slightly older student who just happens to be on the verge of a remarkable podium career. The advice she offers is straightforward, distinctly American in its practicality. There are no purple passages about Mahler's spiritual longing or the pulsations of Beethoven's soul; no priestly advice to cloister oneself in the Rhine Valley. Instead, Marin Alsop is talking about...technology.

"Videotape is an incredible invention for conductors. It's so *alarming.*" She speaks in a husky voice, offsetting her authoritative manner with the calculated self-deprecation of a stand-up comic. "Thanks to video, you can replay all of your movements, all those grunts and grimaces. You find out what is effective and what isn't. You get to see what a fool you're making of yourself. It's startling, but you can really learn from it. If you don't have anybody to videotape you, I suggest practicing in front of a mirror.

"Still, the last thing you should worry about is the gestures, the 1–2–3–4 of it all," she continues. "First, you must learn the piece, inside and out. One of the great things about conducting is that you can practice it anywhere. It's not like the violin or the piano or something. Oh, people might think you're a bit strange sometimes"—she mimics a crazed conductor in wild-eyed, gaping ecstasy—"but you can do *anything* on the New York subway."

The questions are good—specific and intelligent—and she fields them like a past master. (Indeed, Alsop seems most comfortable when in front of an audience; there is a certain reserve in her private conversation.)

"Every orchestra is like this big, bulky, enormous beast," she explains. "It already has a personality. You have to tame it, to work it around to your way of thinking. The conductor is the messenger, the footbridge between the composer and the orchestra. You have to be entirely selfless and entirely self-indulgent at the same time. It's one of those jobs that if you do well, no one pays attention to you, and if you do poorly, everybody complains. What you do is create a visual portrait of the score; the conductor must be able to demonstrate in space what the music should sound like."

At 6:44 she bids the students farewell and moves quickly down the hall to take the Eugene Symphony through Aaron Copland's "Old American Songs." The concert is tomorrow night, and there is a lot left to do. "I work hardest on whatever I'm least good at," she explains as we push our way through guards, stagehands and the occasional well-wisher. "But it's no use, really, because whatever you think will go wrong will go right, while the parts you think are easy will completely fall apart. That's the way it goes."

Everything seems to be going right for Alsop these days. Her career is accelerating madly; she is, to borrow from video terminology, in *fast forward.* Barely two years ago, she was only one among hundreds of young conductors in the country—more gifted than most, to be sure, but still an unknown and likely to remain that way. And then, all but overnight, she won the rapt attention of the music world.

It was in 1988 that she was selected for Leonard Bernstein's conducting seminar at Tanglewood. That happened to be the summer Bernstein turned seventy and "Lennymania" reached unprecedented levels of craziness. Radio and television crews, reporters from magazines and newspapers throughout the world, descended upon western Massachussetts, in vain but Herculean determination to capture every prismatic facet of the Bernstein enigma.

Because Bernstein rarely speaks to the press, the media was forced to follow him from a distance, as he cast his strange glow over concerts and classrooms. And so we watched him work with Alsop, the most obviously gifted conductor in his master class, as she led the Roy Harris Third Symphony—once among the most popular of American orchestral works and a Bernstein favorite. In a performance that was both tender and brilliant, she revitalized the symphony and completely won over her audience. Bernstein embraced her, the visiting critics wrote home excitedly of her work and Alsop became the undisputed young star of the summer.

And then everything exploded. The Eugene Symphony, a good, struggling orchestra in an unusually sophisticated small city, engaged her to be its music director, and she became the associate conductor of the Richmond (Va.) Symphony (a post she has just relinquished). The New York Philharmonic asked her to conduct some concerts in local parks. The Philadelphia Orchestra even put her on its subscription series at the Academy of Music; Alsop was hired on the strength of her press and the Tanglewood video.

Finally, last year, she was hired by the Long Island Philharmonic to take the helm from its founder and director, Christopher Keene.

Of her debut concert with the Philharmonic last February, *Newsday* critic Peter Goodman wrote: "Every phrase was telling, every line had individual coloration, no sound was wasted or used solely for effect... By the finale of the Tchaikovsky, when the brasses rang out nobly and the whole orchestra marched by in triumph, one felt as excited as a kid. Anyone who can revitalize Tchaikovsky like that has got my vote."

Like all successful conductors, Alsop is intensely goal-oriented and impatient with any impediments placed in her path, although she manages to siphon off some of her ambition into a kinetic charm. She is of medium height and sturdy build, handsome rather than beautiful in any traditional sense. Her eyes are large and sensitive, her demeanor calm. She radiates a solid, practical intelligence and seems absolutely uninterested in personal glamour (one adviser told her she needed a nose job; Alsop decided she needed a new adviser).

Still, while Alsop is lively, unpretentious company, one always senses her drive; however expansive she may be at any given moment, her eyes are on the road. Her method of dealing with the world is to *do* things, by force of will, whether or not the odds favor her endeavor. When Juilliard wouldn't let her into its conducting program, she called up some friends and practiced with them instead. Rather than wait for somebody to give her an orchestra, she founded her own. And now— having, it seems, no other option—the world has begun to come to her.

"It's funny, but I have no history whatsoever with the Long Island Philharmonic," Alsop said over a recent dinner in a Eugene restaurant. "I heard them play Beethoven's Ninth on the radio once, and I thought that whoever was handling the microphones was out of his mind. All I could hear was tympani. That was my only impression. But I knew some musicians who regularly trekked out to the Island, and they gave me good reports on the orchestra's progress.

"When I finally went to hear the Philharmonic after I was appointed music director, I found the playing really uninspired. It wasn't Chris Keene's fault; it wasn't really anybody's fault, I guess. Just real dull playing, something very sleepy about it all. I wanted to get up and shake everybody, even the audience. You can't just play the notes! You have to play the *music.* And that's much more difficult.

"I want to turn the Long Island Philharmonic into the best orchestra it can possibly be," she continued. "Now, I know it's made up of free-lance musicians but, believe me, I've been a freelance musician myself, and it *is* possible to inspire them. If a conductor comes onstage 200 percent prepared and completely committed, somehow that's contagious and the players give their all."

Alsop certainly knows the freelance life; she was born into it. Her parents are musicians (mother Ruth is a cellist, father Lamar a violinist), and they played a wide miscellany of gigs—with the Radio City Music Hall orchestra, Fred Waring and his Pennsylvanians, the Buffalo Philharmonic, among others—before beginning long careers in the New York City Ballet Orchestra.

Marin grew up in Dobbs Ferry and Manhattan. "When you're an only child, your parents tend to take you places," she said. "I heard a lot of important artists, and I must've seen *The Nutcracker* a thousand times." She started piano lessons at age two and played until she was six. "I hated every minute of it. But then I was sent away to summer camp, where they taught me the violin. I took to it, and pretty soon I was practicing five hours a day. My parents wisely just stood back and let me find my own way into music."

After studies at the Juilliard preparatory division, she attended Yale University in the early 1970s. "It was a very confusing time in my life. I had already determined to be a musician, but Yale scattered my attention in other directions. Because I couldn't practice and get straight A's and be perfect at everything I did, I wasn't very pleased with myself. So I left Yale and came to New York."

In Manhattan, she received a scholarship to study violin at Juilliard and quickly established herself as a freelance violinist. "I played a lot of commercials, record dates, that sort of thing." But her principal ambition was always conducting. "I'd say that Bernstein was my inspiration, all along, from childhood," she said. "I saw him conduct—probably at one of the Young People's Concerts—and that was it. I thought he was so cool. I was already getting some flak for moving around too much when I played my violin. Watching Bernstein, it was a relief to see that somebody could be so expressive and not only not get into trouble, but actually win prestige."

And so, almost surreptitiously, Alsop began to teach herself conducting. "I would watch every conductor I played under, not only to follow his beat, but also to find out his secrets. If I liked his work a lot, I'd ask for a quick lesson. I worked with Carl Bamberger—a wonderful old man, so musical. Karl Richter gave me a lesson. Later, I studied with Gustav Meier, Harold Farberman and others. I really admired Walter Hendl. While he was conducting up at Chautauqua I would save the money I was earning as a freelancer and fly up to Buffalo—those were the days of People Express—and work with him on the weekends."

But when she applied to Juilliard for post-graduate conducting studies, she was rebuffed. A decade later, she is still angry. "They sent me a form letter back saying that my academic credentials did not meet their standards. Now I had been at Juilliard since I was eight, I studied at Yale, and then I came back and received my master's degree from Juilliard. I can't believe they read my application."

She continued to play her violin—a concert with the Y Chamber Symphony this week, a Billy Joel record session the next—and founded an all-woman jazz band called String Fever, which exists to this day. "I was making good money but growing very frustrated. I needed an outlet to conduct, and my friends weren't coming over for dinner anymore. So I decided to follow through on a daydream of mine, save my freelance money and found my own orchestra."

The result was Concordia—"A Chamber Symphony with a Touch of Jazz," as its early publicity described it. Concordia played its first concert at Symphony Space in Manhattan in December 1984. "That one evening cost me $10,000—all I had—but it was worth it. I got to lead Mozart's 29th Symphony, Stravinsky's 'Dumbarton Oaks,' the Bartok Divertimento and some jazz works. It was a beginning."

Alsop eventually found some financial backers and Concordia settled into a regular niche at Alice Tully Hall, where it now plays several concerts a year. She is obviously comfortable with syncopated, strongly rhythmic material. Unlike most "serious" conductors, who devolve into stiff, campy self-parody when conducting jazz, Alsop always follows through. And she is committed to the propagation of new works.

"Contemporary music, no matter how complicated, has always been very accessible to me," she said. "It's the standard repertory that presents difficulties. But I'm making some headway. I think that I've begun to understand Dvorak and Tchaikovsky. Matter of fact, I've been reading about Tchaikovsky. He had this intense love for gardening but no real gift for it, no control. I think that's a key to his symphonies: He gets a mad passion for something and then it takes over. For the most part, his symphonies are pretty poorly constructed, but there are these wonderful moments, these insights that make them worthwhile.

"Beethoven and Brahms still scare me a bit—haven't found their phone number, wrong area code or something. I'll figure them out. Baroque music is rather terrifying just now; all those conductor-specialists frighten me off. I have an empathy for Haydn because of his sense of humor. He's often very silly, and you have to have some of that spirit yourself when you conduct him.

"I love American music. It is straightforward, even naive sometimes. Even our most complicated music retains a certain earnestness that European music doesn't always have. What we say is what we mean, and we don't have to hide behind needless complexities to get it across."

Alsop wants to conduct in Europe. "There are three strikes against me—my age, my sex and my nationality. I don't think Berlin or Vienna [staunchly conservative, overwhelmingly male-dominated orchestras] will be calling me too soon. But, in America at least, I've found very little resistance to the idea of a woman conductor. It's still unusual enough that the orchestra might even get some publicity for engaging me."

The youth issue is another matter. Unquestionably talented young conductors such as Michael Tilson Thomas and Hugh Wolff have developed reputations for alienating orchestras. "Well, I don't think I'm a *Wunderkind*," Alsop said. "The two gentlemen you mentioned believe wholeheartedly that they are indeed prodigies and that they have something to tell these people who have been at their profession for fifty years. I would never presume that I'm going to tell the musicians anything they've never heard before. The worst thing a conductor can do is assume that he's teaching the players. You're engaged to lead a concert, not to teach school. Even when it's only a semiprofessional orchestra, you must treat the players with tremendous respect."

The respect is reciprocated by the musicians who have worked with Alsop. "Marin is great to work with," Lois Martin, the principal violist in Concordia, said. "She has a great ear, and she rehearses thoroughly and efficiently. She is a better conductor every time I work with her—and not just a little better but a lot."

Kay Stern, Concordia's concertmaster, agrees. "Marin is wonderfully communicative," she said. "She gets results immediately yet keeps everybody feeling comfortable, eager to play their best. It always feels like a real team effort."

Other musicians acknowledge Alsop's technical abilities but admit to some reservations about her interpretations. "She is best at complicated music, with lots of changes in meter and lots of technical difficulties, which she handles perfectly, immaculately," said one player who declined to be identified. "I don't know whether she is a profound musician just yet. Put it this way—there are a lot of conductors who can do more with a Beethoven slow movement. Still, Marin is young and hungry and ambitious and always studying. Sometimes I think the sky's the limit."

Alsop lives in midtown Manhattan but will not discuss her personal life (when the subject is broached, she locks eyes with her interlocutor, smiles firmly, draws an imaginary line with a flattened palm and changes the subject). She would much rather talk about music and her plans for the Long Island Philharmonic.

"Its proximity to New York has been a mixed blessing for the Philharmonic," she said. "On the one hand, you have these wonderful players to choose from, only a train ride away. But the Philharmonic's efforts have been overshadowed by New York and the great orchestras there. Things are changing, though. When I went to Carnegie to hear the Philadelphia a few weeks ago, three different people came up to me to tell me what a great orchestra we were building on Long Island. I said, I know, I know.

"The support of the Philharmonic musicians has meant a great deal to me," she continued. "They got me the job, really. I wasn't even in the pool, and then some musicians who had played with me in Concordia suggested that the board take a look at me.

"Now I want to change the Philharmonic's image. I don't want it to be just that orchestra that plays the Tilles Center. I think it should be a touring group, part of the cultural life of the community. We should play throughout the Island, in community centers, tents, schools, what-have-you. Maybe just pull into town on a flat-bed truck and play there. I'm not really concerned about the trappings as long as the music gets across."

The music is getting across and so is Alsop. In 1989, she made debuts with the National Symphony in Washington, the Louisville Orchestra and the New World Symphony in Miami. This year, in addition to her appearances with New York and Philadelphia, she will conduct the Boston Pops, the Buffalo Philharmonic and the San Jose Symphony. And who can tell where 1991 will find her?

"You see, if you have some perseverance, you can do anything," she said. "Four years ago, nobody would give me a second look as a conductor. My first review was an absolute slam—John Rockwell *killed* me. I didn't get out of bed for ten days. But these things build character. You say, O.K., he's probably a little bit right—damn it—and then go back and look at the video again.

"But I try not to take anything that happens too much to heart, positive or negative. It's all subject to change."

Newsday (1990)

2 A CONVERSATION WITH MILTON BABBITT

This interview was recorded at WNYC-FM studios in 1986 as part of a seventieth birthday tribute to the American composer Milton Babbitt. It has been edited and slightly rearranged for publication.

PAGE: It is always a pleasure to welcome Milton Babbitt to WNYC.

BABBITT: It's a great joy to be here again. I wish I hadn't been reminded of my impending 70th birthday, but I suppose there's no avoiding it. However, I shan't pretend that I like it.

PAGE: I suppose it must be a bit difficult when you reach a certain age and glory and all the celebrations start coming in five-year increments...

BABBITT: ...And one doesn't know if the next five-year increment will happen or not! No, I just wish that I were fifty years old and entering the computer age. I came along a little too late for computers. In 1957 or 1958, Bell Labs called me up and said they were prepared to go ahead with a computer sound production program and would I please join them? So I went down and talked with them, but everything was in a very primitive and clumsy form. They said they wanted me in on it, that they wanted a composer who could understand their mathematics and machine language. And I had just committed myself quite happily to the Mark II RCA Synthesizer and I pointed out that I couldn't possibly do both, so I turned them over to a young composer who was far more mathematically equipped than I. I have therefore just abstained from knowing about the computer, except very passively and descriptively. But I envy those who are working with it.

PAGE: There is a big difference between electronic and computer music, but one finds this recurrent tendency to lump them together and I think both frighten a lot of audiences. That brings us to something we should address briefly because I think far too much has been made of it. You wrote an essay called "The Composer As Specialist" in the late 1950s that was then published in *High Fidelity* magazine and titled, by them, to your everlasting consternation, "Who Cares If You Listen?" Unfortunately—and I know you are as angry about this as anybody—this title has come to represent your public image, insofar as people *have* a public image of living American composers.

BABBITT: I'm glad you added the qualification. No doubt, I'm best known as the writer of this cantankerous little essay. It was published without my knowledge—and certainly without my assent—under the title "Who Cares If You Listen?" The editor explained later that "The Composer As Specialist" was such a dull, academic-sounding title and that this was much more provocative. And when I complained that the title reflected little of the letter and *absolutely nothing* of the spirit of the article—I care very *much* if you listen and *where* you listen and *how* you listen—the answer was something along the lines of "Look, this really doesn't matter very much and it gets people to read the article and that's the important thing."

But it got very few people to read beyond the title and then, if they did read it, they read it in light of the title. I had to do something about it, of course, and some twenty years later *High Fidelity* allowed me to "retract" the title in a taped interview for the magazine. But nobody paid much attention to it, especially since there were new errors throughout. And so I've simply given up and allowed myself to stand as the person who did *not* write an essay called "Who Cares If You Listen?"

PAGE: Let's summarize the essay quickly for, its title aside, it was a significant piece.

BABBITT: Let me tell you something about its genesis. When I was teaching at Tanglewood in 1957, Aaron Copland, who was then the director of the Berkshire Music Center, decided we should have a long, rather involved and rather exhaustive series of lectures covering contemporary composers and contemporary issues for the general public on Friday afternoons. The audience had gathered to hear the Boston Symphony Orchestra and I gave the lecture pretty much off the cuff. I wasn't even sure of what I had said, but it turned out that it had all been taped. There was a man living nearby in Great Barrington who had come to hear me and who liked the idea of printing my lecture. I said no, because I didn't even remember what I had talked about. But he got hold of the tape and sent me a rough transcript and while I didn't completely rewrite it, I certainly revised it and said that it could be published.

My original idea had been to tell the general public what it was like to be a composer—or at least to be my kind of composer. Call it a personal document. I simply had to face the fact that we had a tiny audience for our music, one made up mainly of professionals. I even made a great point out of the fact that we *were* trying to increase the audience but not by sensationalism, not by misrepresenting the music, not by trying to make of it something other than what it was but by attempting to interest listeners in the music as music. Back in the early 50s, we thought we could appeal to what might have been described as our fellow intellectuals with our words and this would bring them around to the sound of our music. But we discovered that what was taken even more resentfully than taking music seriously was *talking* about music seriously. The resentment that this induced made me very, very wary to talk about music again to a general audience.

When I gave the Tanglewood lecture, I made comparisons to other fields—comparing composers to physicists, for instance. Now I wasn't implying for a moment that music is like physics; of course it isn't and I dislike such analogies. But I simply tried to make it clear that music and musicians at this particular moment had problems—severe problems—with regard not only to their audience but with regard to themselves and to performers. Music had changed in fundamental ways which should not be minimized by anyone who would invoke the apparently leveling effects of, if you wish, historical perspective.

I knew well that the music that we regarded as the fundamental music of the 1930s—the music of Arnold Schoenberg, Igor Stravinsky, Roger Sessions, many others—had not become repertory music. We were once told, "Look, you just wait. In twenty, twenty-five years the music that you regard as so fundamentally different, so revolutionary, will either die completely—as an illusion without a future—or it will drop into the repertory and people will be listening to it in Carnegie Hall and it will simply be taken for granted."

Neither event had come to pass. This music had become the cornerstone of the knowledge and thinking of young composers, but it was not being played by the orchestras that moved in and out of Carnegie Hall. That was the basis for my article. I hoped to indicate that this is the way we were living, that these were the facts of our life, and that we were not particularly happy with these facts, that we'd love to have our music published, and recorded, and listened to, and understood, and eventually, if you wish, *loved*—and I do not hesitate to use that word. But this was apparently not the effect of an article entitled "Who Cares If You Listen?"

PAGE: So Milton Babbitt cares if you listen.

BABBITT: But of course! How can one not? What we are interested in when we write music is the listening to that music, the aural effect of that music. And when I say "aural effect" I do not mean titillations of the sonic surface but those works that we think about, that we hope to hear accurately and we hope, therefore, to have played accurately.

PAGE: It does seem that some program notes tend to scare people off. I understand why you write them, certainly, because you are talking about a discipline and you are trying to communicate with your peers and your colleagues. But listeners figure if they can't hear a retrograde inversion, somehow, that there's something wrong with their ears and they run away.

BABBITT: I'm glad you used that example, because a retrograde inversion is the easiest thing in the world to hear. It's simply a repetition of the intervals and if you can't hear intervals in music then you can't hear any music and the next time you whistle Mozart's G minor Symphony, it had better be whistled in G minor or you won't recognize it. But of course we all do recognize the symphony when somebody whistles it in C-sharp minor because the intervallic succession remains the same.

This problem of program notes is one that concerns us constantly. Some of my colleagues absolutely refuse to write them anymore, except to say when the piece was written, the instruments for which it is scored, how many movements it has—and that's it. I understand that approach, too, because one doesn't really know what to say. For example, I'm very concerned never to write anything that will dull the first performance—in other words, that will dull, the singular aural effect of that first encounter.

How does a composer walk that line? Well, you try to provide some kind of information that might have been valuable information secured on a first hearing and treat the upcoming performance as something of a *second* hearing. This is very hard to do. After all, nobody pretends for a moment that words can substitute for the aural experience. So what we do is indicate how the work came into being, rather than what it is doing there or why these things are functioning the way they are in the piece. It's much easier to describe "how" than to describe "why." We know we're talking as the composer and not as the auditor and when we move from the role of composer to that of hearer, we know that this is a very severe change in our relationship with the piece.

So the program note problem is practically insoluble. We can write for our technical magazines, of course, because there people are concerned with the "how" and they understand that this is not a substitute for the "why" or for the reception of the piece. For a composer to prepare a listener, it may be best to indicate not the originalities of the work, nor the singularities that you think justify your work's existence, but rather where it came from, even if the traditions are relatively recent and if the manifestations of those traditions are far below the surface of the music. The trouble then is that they begin to look for all sorts of communal commonplaces instead of listening to the piece as an individual and individuated work.

PAGE: And then the critics try to fit it into some category.

BABBITT: Precisely. And there is nothing I dislike more than talking about "kinds" of music where the piece becomes some sort of statistical sample from the musical population.

PAGE: You're still teaching composition at the Juilliard School, but you retired from Princeton University, where you've spent most of your career and where you continue to live. I find that people who have been associated a long time with universities become sort of honorary figures about the town. I know that Roger Sessions retired from Princeton many years ago but was really considered very much a part of the community until his death.

BABBITT: Oh yes. Roger was indeed a figure in Princeton; he was a figure in many towns but above all it was Princeton. He actually always regarded himself as a New Englander. In 1937, he sent out a *curriculum vitae* when he was desperately looking for a university job and it began the following way: "Born in Brooklyn in 1896, long New England ancestry..." He never forgave the world for having him born in Brooklyn. He wanted to be known as a New Englander but he became a Princetonian.

PAGE: You began your own career in jazz and popular music. How influential was Sessions in steering you into what is sometimes referred to as "serious music"?

BABBITT: Well, that really wasn't the way it went. I don't think that Roger could have steered me much because he had absolutely no sympathy for or understanding of popular music and jazz. You must remember that I grew up in Jackson, Mississippi. That I was born in Philadelphia was one of those accidents like Roger having been born in Brooklyn. In fact, I stayed in Philadelphia for a far shorter time than he stayed in Brooklyn—only a couple of weeks. My mother went there to have me because she believed that only in her own home town—Philadelphia—could her child be properly born. I grew up entirely—to the extent that I grew up at all—in Jackson. So I learned popular music and so-called serious music all at the same time. I began studying the violin at the age of four and then discovered that playing the violin didn't get in the world what playing, say, the clarinet did. I really wanted to play the trumpet, actually, but they made me play the clarinet.

I was playing gigs by the time I was ten years old with New Orleans jazz men, but all the time I was very much aware of what was called serious music. At the age of ten, I traveled to Philadelphia with my mother and an uncle of mine who was then a student at Curtis. My uncle and aunt and people like Marc Blitzstein and Isidor Fried were all involved in playing contemporary music, talking about it, arguing about it. And that's when I heard Schoenberg's music for the first time.

I don't remember if it was Opus 11 or Opus 19 but I became very aware that there were these two separate worlds of music. And they've always been utterly disjointed for me. I never thought there was much of a relationship between the two, except that I began to recognize in some popular songs certain nonrepetitional devices that I identified with techniques of serious composers. (I am always amazed by Irving Berlin's "All Alone"—an incredibly sophisticated song!)

From that point on, I never thought of getting into popular music seriously. In fact, I didn't really want to get into music seriously either; I'd seen too many tragedies already. I decided that I couldn't fight this battle but, of course, I ended up trying to do it anyway. I came to New York, and got so involved with serious music that when I graduated from New York University in 1935, I went right to Roger Sessions for private study.

He had absolutely no knowledge of popular music but I'll tell you an amusing anecdote. One day, Sessions was with Artur Schnabel and they had been reading *The Nation* or *The New Republic* or some such magazine and there was a reference to a popular, corny tune called "The Music Goes Round." And Roger said "Milton, you know something about popular music, don't you?" And I said, sure. And he said, "Do you know this song?" And I said that I didn't really keep up with these things anymore but, yes, I certainly knew "The Music Goes Round." And he asked me to play it. So I went to the piano—I'm sort of a cocktail pianist—and I played a little bit of "The Music Goes Round." So Schnabel says, "You twisted your wrists." And I apologized: "You know, I'm not really a pianist, Mr. Schnabel; I just sort of picked this up. I've never been instructed on the piano." And he said: "Yes, well you must never twist your wrists, you must bring your hands over firmly like *this*." And so now I may say that I've studied with Schnabel!

PAGE: How do you feel listening to your works from forty years ago? Do you ever feel the urge to pull a Paul Hindemith and rewrite them completely?

BABBITT: I can't imagine "pulling a Paul Hindemith" in any respect. But of course I know what you mean. My early "Three Compositions for Piano," which date from 1947, I never wanted to rewrite at all, even though the only person who could play them back in the late 40s was Robert Helps. It wasn't until very recently that anybody else could play the pieces—particularly the first one—up to tempo and that person is Robert Taub, who has also recorded all of my piano music. So I'm happy to hear those pieces finally played with the fluency and ease which Taub brings to them.

Of course, I regard some of my early pieces as works that I would not write now. I regard them not so much as simple, nor so much as rudimentary, but as "youthful" pieces. For example, my "Composition for Four Instruments," which was the first piece of mine to be published, is a piece that I'm constantly having to look at in classes. People ask me about it, students are playing it, a group at the New England Conservatory has even recorded it. But I can only look at those pieces with the realization that I haven't changed very much. Now some may regard this not as evidence of compositional probity but just sheer intransigence and stubborness, but I'm perfectly willing to stand on those pieces.

Please remember that we're talking only about compositions written after World War II, after I was thirty. The pieces before this time I have retired. There was a piece that won the Bearns Prize, "Music for the Mass," which you can hear only by going to the Columbia University Library and looking at the score and listening to it internally. There was a string trio that Gunther Schuller constantly threatens to perform but now that he's no longer at Tanglewood he probably won't. And there's not really very much else from that period that I'd be interested in having anybody hear. I'm more interested in what happens next.

3 MARIA BACHMANN: THE MAKING OF A MUSICIAN

Music is a glorious art and a difficult profession. And there are few rites of passage so important to the profession—if not necessarily the art—as the debut concert.

Into Manhattan they come—child prodigies and elderly dreamers; self-trained mavericks and tenured chairs of departments; brilliant conservatory graduates and wealthy dilletantes—to proclaim their formal entry into the music world. An auditorium is engaged, most likely Weill, Merkin, Town, Tully or (for the truly wealthy or truly grandiose) Carnegie Hall. A photograph is chosen, a flyer prepared, press releases are dispatched to newspapers and radio stations. Finally the hour arrives; the house is peopled with family and friends, a couple of curious ticket-holders, perhaps a representative or two from the music business and, it is hoped, a critic to tell all about it.

As it happened, I attended more than one hundred debut concerts during my apprentice years as a critic. Most of them have long passed from memory, a melodious blur of entrances and curtain calls, duly recorded, appraised and forgotten. The level of technical competence was generally high, the level of musical inspiration rather less so, and it often became my duty to write a review in mezzo-forte—a notice that would give credit where it was due, express reservations honestly but with a certain gentility.

Still, there were those few concerts that transformed the debut shift, for a moment, into the most exciting on the beat, when one was presented with an extraordinary artist and had the opportunity to break the news:

> Maria Bachmann is a violinist of soul and patrician refinement; her beautiful debut recital at Town Hall Tuesday night was one of the most rewarding of the season... Ms. Bachmann has the strength, agility and precision that we have come to expect from a modern violinist, yet her playing has an affecting warmth that is distinctly unusual. By "warmth" I do not mean effusion, for Ms. Bachmann's playing never tears passion

to tatters, and there is evidence of a cool intelligence behind her interpretations...
  The New York Times, March 1, 1987

  I met Maria Bachmann some three and a half years after those
words saw print, on a fine, cold evening last October in Muskegon,
Mich. In the interim, I had followed her career with more than usual
interest, heard her several times and never been disappointed. Here was
a violinist equally at home with the standard repertory, whether Brahms,
Mozart, Beethoven or Kreisler; with the recondite solo sonata of Bela
Bartok; and with the most challenging contemporary music.
  The response to Bachmann's work throughout the country has been
similar to my own. "From the very first bars, it was clear that the young
violinist has the expressive capacity, the poise and the technical where-
withal to be placed among a very select group of musicians," the *Los
Angeles Times* wrote of her West Coast debut. The *Boston Globe* called
her "astonishing in every musical and technical regard." And meanwhile,
her premiere recording has been issued on the Connoisseur Society
label; it contains Beethoven's "Kreutzer" Sonata and the ambitious four-
movement Sonata that George Rochberg composed especially for her.
  In an earlier generation, when the galaxy was larger, Maria Bachmann would likely have been a star by now. All the traditional ingredients
are there: a unique talent, critical hosannas, personal charisma, the
willingness and the ability to work hard and long. Yet with the constant—
and seemingly irreversible—diminishment of the audience for classical
music in this country, expectations have changed.
  If the audience is both aging and dwindling, it is also increasingly
obsessed with the past. The world's most famous tenor, seventy years
after his death, remains Enrico Caruso (this, despite the energies of
Luciano Pavarotti's handlers). The superstar pianists, still alive on compact disc and videotape, are Vladimir Horowitz, Artur Rubinstein and
Glenn Gould. And, now that Leonard Bernstein is gone, I have found
that many college students can no longer name a living conductor.
  A handful of artists are familiar to the general public but often for
extramusical reasons—Isaac Stern playing in Tel Aviv as the Scuds fall,
Itzhak Perlman telling a joke on the "Tonight" show. But more and
more, it sometimes seems, the serious efforts of classical musicians will
occupy a niche similar to that now occupied by professional philosophers
and poets—honorable company, to be sure, but a finite, specific elite,
with all the ramifications the distinction implies.
  And yet...

Maria Bachmann's name is high on the marquee at Muskegon's Frauenthal Center, and my cab driver informs me, with genuine excitement, that he has two tickets to hear her play with the West Shore Symphony on Saturday night. The Frauenthal Center, with its elaborate art deco *cum* "Arabian Nights" interior, is one of those RKO palaces from the silent era, claimed by civic groups just ahead of the wrecking ball. The building also contains a Christian Science Reading Room, an upscale bar/restaurant and a bowling alley.

Muskegon itself is pure Dreiser—smokestacks and open skies, freight yards and mansions gone to seed. The inner city is severely depressed (with the highest unemployment rate in Michigan after fabled Flint). Industry is all: The Frauenthal Center "angels"—corporate supporters who keep the endeavor alive—include American Grease Stick and the Lake Welding Supply Co. Within two weeks of Bachmann's visit, the theater will present a touring ballet company, a local staging of *Romeo and Juliet* and a concert of gospel music. "This is not what you'd call a culturally impoverished city by any means," the driver says proudly as I pay my fare.

Bachmann arrives for rehearsal unescorted and unannounced, takes her place onstage and shyly greets the orchestra. She is thin and graceful, dressed in heels, green khakis and a white silk shirt, and looks about a decade younger than her thirty years. She removes her violin, tunes it, shakes some broken hairs from her bow with a look of mild irritation and confers briefly with West Shore music director Murray Gross. Then he lifts his baton, and the orchestra begins the Tchaikovsky violin concerto.

She listens to the short introduction, nodding her head with approving camaraderie. As she begins her solo, in deepest concentration, she closes her eyes, but always opens them again just at that moment when it is time to take a cue from the podium. During the inevitable minor points of disputation that come up between conductor and soloist, Bachmann's manner is polite but firm. She expresses her wishes with tact and delicacy, but she *does* express them, and they are heeded.

Bachmann, a cool, clean violinist, eschews histrionic excess. Her tone—lean, centered and atypically dark-hued—may initially seem austere when compared with the lush, effusive outpouring favored by her contemporaries. She makes no attempt to drown the listener in pretty sound; rather, one is drawn in by the elegance of her phrasing, the clarity of her musical thought, her ability to sustain a musical narrative. And yet she gives the Tchaikovsky its full, sweet, gypsy warmth. When the rehearsal is over, the orchestra applauds.

As does the audience, wildly and without hesitation, the next evening, immediately after the first movement of the Tchaikovsky—a lapse of etiquette that is often rewarded with a frozen and uncomfortable stare from the stage. Instead, Bachmann acknowledges the ovation briefly and gently, makes it clear that her work is not yet done and prepares to begin the second movement. When the last notes of the concerto die away, the audience immediately rises to its feet. The following day's paper confirms what one already knows: Bachmann has met Muskegon, and it is hers.

"The fall was chaos," Bachmann said over tea during a midseason respite, in a bright voice that is melodic and almost childlike. "There was the Tchaikovsky in Muskegon, a week's residency in Iowa—three appearances a day: nursing homes, public schools, hospitals and then a formal recital—and a concert upstate. Then I flew to Stockholm to learn some contemporary Swedish music for a concert in Denmark but came down with the chicken pox—the *chicken pox,* can you believe it?—and upon the advice of my doctor and my folks, I had to cancel.

"Now, I *never* cancel concerts. I'd only missed one in my career; I've played with strep throat, flu, fever, everything. But I'd done seven hours of rehearsal every day for three days, getting sicker and sicker, and then I was expected to take an overnight train to Denmark for the show. And I couldn't do it, you know?

"The sponsors were not very understanding. They had put me up in a four-star hotel, and when I went to check out, I found that they'd refused to pay my bill. I lost my fee, of course, and then because I had a nonrefundable ticket and wanted to fly home early, I had to buy another one—$1,200 or something. And then I came home and went straight to the hospital, X-rays, I.V. and everything. All because of a bug I picked up running around Iowa.

"And there's no insurance plan to cover a solo artist, no unemployment compensation—despite the fact that I picked this illness up on the job, despite the fact that I'd already learned the music, done the rehearsals, done everything but play the concert. So I came back to the States some $4,000 poorer."

Ah, life on the road. . . . Other recent adventures include a performance in Colorado Springs on the same day that a plane fell into the foothills nearby; a visit to Moscow for the Tchaikovsky Competition, where she placed seventh ("We couldn't find food; the contestants were

literally hungry all the time!") and—to counterbalance this grim picture—grand moments such as her two concerts last October in Hungary,
whence her parents had fled some thirty-four years before.

At the time of the foiled 1956 revolution, Maria Bachmann's father,
Tibor Bachmann, an outspoken anti-Communist, was in a Hungarian
prison. "My mother basically bribed a prison guard with some jewelry
that she had, and they let him escape," she said. "They got to a village
near the Austrian border and loaded into a big hay wagon. At this time,
the guards were pretty lax about letting the local peasants cross over
and visit their relatives, and so the cart was waved right on through. If
my parents had said anything, it would have been obvious that they were
city people, and they would probably have been arrested."

Tibor and Eva Bachmann arrived in Vienna with a four-year-old
son and went directly to the American embassy; two months later they
were in Chester, Pa., a small city in suburban Philadelphia, where Maria
was born in 1960. "My father was a musician and musicologist—he'd
been an impresario in Hungary. My mother was trained as a nurse. First
we lived in a tiny house in Chester, and then we worked and saved and
moved to a larger house in Chester, this one with a studio for my father.
And so when I was a baby I was listening to my dad's students all day,
coming in and out, playing through their stuff. It must have made an
impression: When I was two or three, I got up at about six in the morning
and picked out a Bach minuet on the piano. And so they started me on
lessons."

But the lessons didn't take. "I didn't like to practice, and so I was
always making mistakes, and my dad would become cross with me," she
remembered. "But one day when I was seven, I found a violin in our
basement. I didn't know what it was—at the time, I thought it was a
guitar, actually—but it caught my interest, and I began lessons. Well,
learning two instruments got to be a little much for an eight-year-old,
so I gave up the piano and put all my efforts into the violin."

The efforts paid off quickly. When she was eleven she played her
first concert; by the time she was in her teens she was presenting a
recital every year (the program might include a Bach solo partita, the
Brahms Sonata in A, and Stravinsky's "Suite Italienne"—hardly kid
stuff, in other words). Tibor Bachmann was now a professor at Indiana
University of Pennsylvania, one hour outside of Pittsburgh, and Maria
joined the Pittsburgh Youth Symphony, where she was assistant concertmaster for three years. She won local fame as the town prodigy.

At this point, Bachmann had no plans to become a professional violinist. "I liked music, but I still didn't enjoy practicing. I thought I'd become an architect or a clothes designer or maybe a painter." However, when she was seventeen, Bachmann was admitted to the Curtis Institute of Music in Philadelphia, a small, prestigious conservatory that charges no tuition and has produced an extraordinary number of distinguished artists. There she studied with Ivan Galamian (who also taught, among others, Michael Rabin, Pinchas Zukerman, Kyung-Wha Chung and Perlman) and, after Galamian's death in 1981, with Szymon Goldberg.

"Goldberg is a masterful violinist whose sole concern is the interpretation of great music, to the exclusion of all virtuoso frills," the late Boris Schwarz wrote in his invaluable book *Great Masters of the Violin*. "His technique is flawless, his tone warm and pure, his sense of style and his musical taste exquisite. His performance style stresses refinement, intimacy and noble intensity, equally evident in the classical repertory and modern works."

In Bachmann, Goldberg had found a pupil whose creative temperament was in perfect sympathy with his own. "He greatly opened my options as a musician," she told *Newsday*'s Peter Goodman in 1989. "I felt very much prepared after I left Curtis to pick up a new piece and search for some sort of truth in the score rather than just playing it . . . He found meaning behind what was written on the page."

For better and for worse, competitions are a familiar—if controversial—part of the music world; there is no doubt that a major prize can help the career of any young artist. In 1983, Bachmann won the Fritz Kreisler International Violin Competition in Vienna, and three years later she took both the first prize and the U.S. Trust Artist Award of the Concert Artists Guild New York Competition, which funded her debut.

Concert Artists Guild is a 40-year-old nonprofit organization devoted to fostering and nurturing the development of young musicians; it has continued to represent Bachmann professionally. "They have been wonderful," she said, "and I appreciate their philosophy—which is to keep their artists busy on a lot of different levels. I know of some other groups that charge higher fees for their musicians, but I don't think they play as often. Concert Artists Guild believes in keeping you *out there*, keeping you working; they'll go out of their way to make sure the concert takes place. And you learn a lot by playing a lot."

She now plays forty to fifty concerts a year. "And I'll admit that there are occasions when I feel disappointed after the show is over. For instance, Muskegon: I didn't think that was such a terrific concert. I would have liked to have done it better. But it's really hard to give your absolutely most inspired performance every time, and I think it gets harder the more concerts you give. So it's a trade-off, I guess."

Although she is generally open and animated in conversation and quickly names conductor Sergiu Celibidache, cellist Yo-Yo Ma and pianist Alicia de Larrocha as three of her favorite musicians, Bachmann seemed reluctant to discuss any of her violinist colleagues. "My two favorites are both dead, unfortunately—David Oistrakh and Jascha Heifetz. Oistrakh was my first hero; I heard him play with the Philadelphia Orchestra when I was about seven, and he made a tremendous impression on me. I loved his sound so much. But I think Heifetz has an even greater musical integrity; I've now realized that Oistrakh had a fair amount of technical problems, and he sometimes altered his interpretations to ease some of his difficulties. Whereas, for Heifetz, there *were* no difficulties.

"And then there were Kreisler and Szigeti. There was a kind of freedom and expressiveness about the way they played that you don't find anymore. Musicians are so afraid to be free, I think, because there is that tremendous concern for technical perfection. It's a result of the recording industry. Everything you hear on record is perfect; you can play a piece over and over again and splice it until it's perfect. So when an audience comes to a concert, the first thing they want is technical perfection. And with most of those violinists—even Kreisler—the playing is sometimes not quite in tune. Beautiful and inspired and soulful, yes, but not always entirely accurate. And I wonder how that would go over with an audience today."

Bachmann plays a modern violin by Sergio Peresson, an instrument maker born and trained in Italy who is now based in Haddonfield, N.J. "I know some people whose families have taken out all their life savings and mortgaged the house so that they can play on a Strad. Look, if you can afford it and you want to pay those kind of sums, why not? But there are great instruments being made today; I know because I play on one of them. I visited Mr. Peresson's house, and I picked up this fiddle and knew it was right for me immediately. I touched the open strings and *knew*—not just its volume and projection but the beauty and intensity of the sound."

She now divides her appearances between recitals (usually with Jon Klibonoff, a strong, skillful pianist in whom the late Glenn Gould took a particular interest); solo performances with orchestra (she has about 25 concertos in her repertory) and chamber music. Throughout the summer, she is a featured artist at the Bowdoin Summer Music Festival in Maine; her boyfriend of many years, Alexander Simionescu, with whom she lives in Paramus, N.J., is the violinist with the Bowdoin Trio.

Bachmann moved to Paramus after two years on West End Avenue. "I hated living in New York—too many people, too much traffic, too much crime, too much noise. I'm very sensitive to sound and find that I need a calm environment to get my work done. At the end of the summer I used to come back from Maine such a relaxed, happy person; two weeks in New York and I'd be a wreck again."

The spring itinerary includes a return visit to Hungary; concerts in Paris, Rome, Milan and Turin; several recitals in California (including a private musicale at a home in Santa Monica) and then back to Maine. Throughout it all, she makes an effort to practice between three and four hours a day. "I'm always trying to strengthen my left hand. And, beyond the physical action of playing a violin, it's terribly important to *think* about music—to read it, listen to it, steep yourself in it."

What happens now to Maria Bachmann? It is not only dangerous but impossible to predict the course of a career. She has clearly risen to a certain prominence in her field; she makes a decent living (no small accomplishment for a solo performer in 1991); her work is known to critics, tastemakers and her fellow musicians. But she admits to some anxiety about the future: "The conservatories turn out a lot of good violinists," she said.

Indeed they do—and some of them have powerful friends. With all due respect to the considerable talents of Midori and Anne-Sophie Mutter, it is safe to say that the former owes much of her celebrity to her association with Bernstein, and the latter was a protégé of both Herbert von Karajan and Mstislav Rostropovich. A well-connected mentor could help Bachmann immensely. So could a major record deal or a contract with one of the big management companies. But these might well be Faustian bargains: She would then be primed, packaged, taken firmly in hand and given an "image," while flashy Paganini caprices and other showstoppers replaced Rochberg and Bartok in her repertory.

"Sure, I'd like my fees to go up a bit," Bachmann said. "But, overall, I'm satisfied. Because, finally, it has to be the music itself that keeps you going. I want to keep playing the great music of the past and the best music of the present; I want to play with some more orchestras and make some more recordings. I don't know the really influential people in the business, and I don't think that I'm destined to make huge piles of money, but I can have a long and valuable career if I work at it, and that's more important to me."

Newsday (1991)

4 LEONARD BERNSTEIN AT 70

Lenox, Mass.

This pastoral Berkshire village, which has long nurtured the brilliant, wealthy and eccentric, is preparing to welcome home a man who personifies town values.

On Thursday, Leonard Bernstein (Berkshire Music Center, classes of 1940, 1941; chief of conducting faculty, 1951–55; eminence—gray and otherwise—ever since) will observe his seventieth birthday, and a celebration of gargantuan proportions has been planned.

Tanglewood, the summer home of the Boston Symphony Orchestra, will present a gala concert featuring Lauren Bacall, Van Cliburn, Lukas Foss, Yo-Yo Ma, Bobby McFerrin, Frederica Von Stade and the prodigious young violinist Midori, among others. Beverly Sills will be the host. New works, dedicated to Bernstein, have been composed by Ned Rorem, André Previn, Stephen Sondheim, Michael Tilson Thomas and others.

Some fancy guests are expected. Malcolm Forbes and Elizabeth Taylor are planning to arrive in a hot air balloon. Mstislav Rostropovich will fly from Europe to Kennedy Airport on the Concorde, to be immediately whisked to Tanglewood by limousine. Kitty Dukakis, Joan Kennedy and Kitty Carlisle Hart will be there. Quincy Jones is coming in via helicopter. Hanae Mori has designed an official T-shirt to commemorate the event. Tickets for the gala range from $50 to $5,000, and simple admission to the Tanglewood lawn has been jacked up to $20.

And now a sort of "Lennymania" has swept through Lenox. He is the talk of the pretty, pretentious cafes that line Church Street. A photo gallery at the Lenox library has devoted a section to shots of Bernstein in action. "I can tell that Lenny is in town," Heinz Weissenstein, Tanglewood's official photographer from the beginning half a century ago, said with a grin, motioning to a television truck parked outside the main gate.

Music students, one of whom solemnly informed me that Bernstein was God, are on a frantic hunt for the maestro, who is variously reported to be staying in one of two castlelike mansions turned resorts and in a private house in neighboring Great Barrington. And somewhere, no doubt, a party is going on.

There are always parties around Bernstein. Some of them are polite, society functions, at which he charmingly misbehaves ("Lenny is so rude he takes my breath away," Rorem wrote, after a Bernstein dinner party, in his most recent published diary). More to the point are the dusk-to-dawn bacchanales, in which Bernstein the haunted insomniac holds court, puffing on an endless chain of cigarettes, gulping whiskey and expounding on any subject that enters his mind.

At the so-called harmonic convergence a year ago, it is said that Bernstein outdid himself. At Serenak, the mansion high above Tanglewood that was once the home of Serge Koussevitzky, he set up rows of mystic crystals, watched the sky like a lycanthrope, paced the yard and talked through the night—poetry, pronouncement, premonition—refusing to let anybody in the company precede him to bed.

Last year, Joan Peyser published a controversial biography of Bernstein that alleged, in considerable detail, drinking, insecurities, gratuitous cruelties and promiscuous homosexual affairs. These assertions were hardly news to anybody in the music business; indeed, most people I know were surprised by the number of famous Lenny stories that were left *out* of the Peyser book.

They sell the biography at the Tanglewood Music Store, but it is a strictly under-the-counter item. "Oh, we can help you out with the Peyser book," the smiling teenager behind the cash register affirmed, with a look of complicitous camaraderie. "We have lots of copies. They're just not on display. You know how it is." A friend later informed me that there were "high-level" meetings among the Tanglewood brass to decide what to do about this book, and the decision was made to turn it into the musical equivalent of *samizdat*.

The focus of these rumors, the center of this hysteria is a short, paunchy, distressed man with a magnificent head and the face of an amorous saint, who summarily canceled his initial morning conducting class at Tanglewood ("We won't go into the reasons," he later explained in his husky bass, eyes rolled heavenward) but arrived ten minutes early for an afternoon coaching session of student conductors at Serenak.

There were no mystic crystals in sight this day, just two pianos, a handful of breathless conducting students, journalists and various representatives of the music business packed into every corner of the room. Bernstein entered, dressed in jeans, a pink workshirt and wooden beads, and proceeded to give bearhugs to roughly a quarter of the visitors, shaking hands with others and greeting still more by name.

One is always surprised by the man's height. Bernstein seems a positive Olympian on the podium, and yet he cannot be much more than 5-foot-4. Yet such is his physical presence that he commands attention, and he held the room in thrall.

"How'd it go this morning?" he asked a conducting student.

"OK," came the reply. "We read through Roy Harris' Third Symphony."

"With any luck?"

Timid laughter from the room. "*Some* luck."

Bernstein then sat down and began a brilliant exegesis of the Harris work, once perhaps the most popular symphony written by an American, now in eclipse. He is gentle, specific, inspired with the students. "Here we have a very early example of minimalist music," he said of a passage for strings which did, in fact, sound a lot like Philip Glass.

"One of the hardest things to get from an orchestra is that long legato," he continued. "And a long lean line is what Harris is all about. You want the strings to sound like 12 string quartets with an endless bow; you want the brass to sound like a gigantic choir with endless breath. It must go inevitably, in one line."

"Read the mind of the composer and recompose the piece with him," he told another student who was struggling through the first Essay for Orchestra by Samuel Barber. He listened for a few minutes. "I like the way you conduct," he said. "The way you show dotted rhythms; it's very clear. And you know the piece. You've misnamed some chords, but that's a linguistic problem. You're doing fine.

"*Show* me the melody," he went on. "Visualize it and then put it into your beat. See the shape of the melody and then make the orchestra see it too." Bernstein worked with the student on the Barber for more than half an hour but never proceeded past the first few bars. "Once you can conduct these few measures right, you'll have no trouble with the rest of the piece. It all builds from this."

Watching Bernstein himself conduct today, one notes the unusual combination of different elements that make up his podium manner. His leadership is simultaneously indulgent and impeccably controlled, Dionysian yet carefully thought through. He is Chaplinesque in his pantomime and can convey the essence of a particular measure with such acuity and grace that watching him is like reading a map of the score. When he wants to, Bernstein can be a tremendously economical conductor, with a tight little beat that is scarcely visible yet gets the

necessary results. But, more often, he wriggles and dances like an orgiastic surfer, carried aloft on waves of sound.

Entering his eighth decade, Bernstein shows no signs of slowing
his pace. He has recently recorded Puccini's *La Bohème* in Italy with an
all-American cast of singers. He is preparing to tape the Mahler Sixth
in Vienna. He is in the midst of composing a song cycle called "Arias
and Barcarolles" for two mezzo-sopranos, two baritones and piano, four
hands—a twentieth-century answer to the "Liebeslieder Waltzes" of
Johannes Brahms. And he has been meeting with Martha Graham, who
is now in her nineties, about the possibility of future collaboration on a
ballet.

"I think Lenny can't bear the fact that he is seventy," a friend of
half a century said. "He has so much he wants to say, so much to give.
Most of us have come to terms with growing older. But not Lenny. He
wants it to go on forever."

Newsday (1988)

5 LEONARD BERNSTEIN: IN MEMORIAM

Endowed with protean energy and irrepressible enthusiasm, Leonard Bernstein—composer, conductor, pianist, author and educator—had as profound an effect on classical music in America as any individual.

His long association with the New York Philharmonic, which spanned nearly half a century, is worth an essay in itself. So is the music he wrote for the theater: Bernstein not only composed one of the indisputable masterpieces of the Broadway repertory—*West Side Story*—but several other musicals, operas and an ambitious staged "Mass."

But perhaps Bernstein's most important influence was the example he set for American musicians. We have long suffered from a cultural inferiority complex in this country. When Bernstein began his career, the phrase "good music" was widely used as a synonym for "European music"; moreover, it was believed that only the Old World could provide the proper spiritual and technical training for a young artist. Indeed, to this day, when choosing a director, our major performing arts organizations show a dismaying preference for second-class Europeans over first-class Americans.

However, extraordinary progress has been made, and we have Bernstein to thank for it. He elected to build a career in the United States—studied here, made his debut here, lived here all his life. This choice had something of the same effect on our native musicians that Ralph Waldo Emerson's lecture "The American Scholar" did on nineteenth century *literati*. It was a declaration of independence, nothing less. And, eventually, Europe—indeed, the world—came to Bernstein and, by extension, to America as well. Although the way of a young artist remains a difficult one, no longer do our musicians face the cultural lock-out that was once taken as a matter of course.

American music found an important friend in Bernstein; he conducted music not just by his mentors (Aaron Copland, David Diamond) and friends (Harold Shapero, Ned Rorem), but by creators as diverse—and hard to put across—as Elliott Carter and John Cage. He helped win a new, academic respect for jazz and rock. He was a convinced adherent of stylistic eclecticism and would occasionally program works that he didn't personally like very much, simply because he believed they were serious efforts and deserved to be heard. The resulting performances were invariably idiomatic and assured.

Still, had Bernstein never been the music director of the most prestigious orchestra in America's largest city for more than a decade, had he never written even a note of music, he would occupy a unique place in our annals as the most influential music teacher in history.

A big claim, of course, but one that stands up to analysis. Composer and critic Virgil Thomson called him "the ideal explainer of music, both classical and modern," and, for many years, Bernstein had the medium of television with which to disseminate his ideas. Through "Omnibus," and later through the televised New York Philharmonic "Young People's Concerts," he charged an entire generation with the love of music. Indeed, when one calculates the number of television appearances Bernstein made in his prime, the total number of people he reached must be astronomical.

I am one of Leonard Bernstein's students. So, most likely, are you, if you are between the ages of twenty-five and fifty and care anything about music. For those of us who were raised in rural surroundings, far removed from concert halls and subscription series, watching and listening to Bernstein take us through music on the small screen was like passage into a new world. He gave naturally to the television medium and, with his graceful movements, his poetic face suffering and ecstatic, he seemed to embody the essence of music in all of its pain and exaltation.

Bernstein's own compositions were typically best when at their least determinedly "serious." *West Side Story* is show music at its most direct and invigorating and it has dated hardly a bit in thirty years. The overture to *Candide* remains a masterfully compressed mixture of jaunty, jazzy aggression and swooning romanticism. But this is Bernstein's "show" music, and there is evidence that he did not esteem it so highly as did the rest of us (he steadfastly refused to conduct *West Side Story* in a Broadway pit).

The man himself was complicated and unpredictable—alternately extraordinarily sensitive and appallingly thoughtless. Tom Wolfe's *Radical Chic,* inspired by a party Bernstein's wife, Felicia, threw for the thuggish, anti-Semitic Black Panthers, remains a masterful demolition of the conductor's sociopolitical pretensions. A later biography by Joan Peyser recounted, in some detail, Bernstein's drinking, drug use, sexual promiscuity and less-than-admirable personal traits.

Indeed, it is hard to think of any indisputably great artist since Richard Wagner in which the sublime and the silly were so inextricably

yoked. To complain, as many did, of Bernstein's all-encompassing ego-
ism was to miss the point. He *was* ego, ego personified, which led not
only to cosmic performances of Mahler and Beethoven but to the practice
of publishing doggerel verse, leisure-hour jottings and old school papers
in a glorified coffee-table volume called *Findings*. And those who came
to interview Bernstein often found that specific questions about, say,
Carnegie Hall, were likely to be answered with a long digression on the
bombing of Hiroshima (which, for the record, he thought was a bad
thing).

Composer Ned Rorem once offered this assessment of Bernstein:
"One hesitates to tell him that he is not a Thinker, like the earthbound
Chomsky whom he forever uncritically praises, but an airborne doer
who doesn't need to prove how smart he is."

It is, of course, the "doer" who will be remembered. "It is impossible for me to make an exclusive choice among the various activities
of conducting, symphonic composition, writing for the theater and playing the piano," Bernstein wrote in 1946. "What seems right for me at
any given moment is what I must do . . . For the ends are music itself,
not the conventions of the music business." Bernstein followed this
credo to the end.

I attended his last two concerts, on Aug. 14 and 19, 1990, in Lenox,
Mass. They were events both solemn and uplifting. We knew he was
sick and some of us believed he was dying. But the performances were
transcendent, and I suspect that Bernstein would not have been entirely
displeased that Tanglewood proved the site of his farewell to music.
Half a century ago, he studied conducting there with the legendary
Serge Koussevitzky; in the intervening years, he grew from a brash,
hungry and superbly gifted student into our unquestioned "Grand Old
Man" of music. Yet he returned to Lenox every summer—to instruct
and inspire the young musicians.

His conducting, at those last concerts, was startlingly low-key. For
years, his podium manner had been characterized by a balletic ecstasy—
what Bernstein called his "Lenny Dance"—and he once admitted that
he never knew where he was while in the middle of a performance. At
Tanglewood, there was no Lenny Dance and, for the first time in the
experience of this long-time Bernstein watcher, the orgiastic extrovert
seemed cautious, reined-in, measuring his every motion with gravity and
care. He nearly broke down in the middle of Beethoven's Seventh Symphony and conducted much of the third movement leaning back against
the podium brace, gasping for breath.

When it was over, Bernstein, shaken and spent, was accorded one
of the most rapturous standing ovations I've ever witnessed; even the

critics joined in. Slowly, painfully, Bernstein walked back from the wings to greet the audience. He smiled weakly, made a motion that might have been construed as a benediction, and left the stage. He had fought another battle and, for the last time, he had triumphed.

Newsday (1990)

6 A CONVERSATION WITH JOHN CAGE

John Cage and I spoke in WNYC studios, New York, for a nationally syndicated radio program called Meet the Composer. We began our discussion with the onset of Mr. Cage's career in the mid-1930s.

JOHN CAGE: I remember when I applied for a job with the WPA in San Francisco, the music department told me I wasn't a musician. I said I needed a job and I worked with sound, so where could I go? They told me to try the recreation department. I got a job with a San Francisco hospital, a job keeping children quiet. Since I was interested in sound, they said, I could try to stop the children from making any. So I went to the hospital and taught the children rhythmic games. I had them be very quiet, take their shoes off and walk around the room.

TIM PAGE: But you had already studied composition with Arnold Schoenberg, and *he* certainly knew you were a musician. As a matter of fact, he took you on as a scholarship student.

CAGE: When I asked Schoenberg if he could teach me, he said his price was probably beyond my means. I said don't mention it, because I don't have any money at all. He asked me if I would devote my life to music; I said yes, and he agreed to teach me for nothing.

PAGE: What did you learn from Schoenberg?

CAGE: He was an extraordinary teacher, and I don't know if I can answer your question and do him full justice. Once he sent us to the blackboard to solve a problem in counterpoint. He gave us all the same *cantus firmus:* We knew it by heart because in the two years we'd been working with him, he'd never changed it. It was C-D-F-E-D-C. So I solved the problem; he said I was correct and told me to do another. And then another. Anyway, I gave him about eight or nine solutions and he continued to ask for more. Finally I said—

not at all sure of myself—that there weren't any more
solutions. He told me I was correct. Then he asked what
the principle underlying all of the solutions was. I couldn't
answer. I had always worshipped Schoenberg as though
he were superior to other human beings, but at this moment
he seemed so vastly superior that I was speechless. This
happened in 1935, and it would be at least 15 more years
before I could answer his question. Now I would answer
that the principle underlying all of our solutions is the
question we ask.

PAGE: Schoenberg later characterized you as an inventor of genius
and you rose to prominence with some intriguing works
for bones, rattles, specially designed percussion instruments and the prepared piano.

CAGE: Well, I had become convinced that everything has a spirit
and that everything sounds. I became so curious about the
world in which I lived, from a sonic point of view, that I
began hitting and rubbing everything I came near—
whether I was in the kitchen or outdoors, and I gradually assembled a large collection of unconventional instruments.

PAGE: How do you prepare a piano?

CAGE: If you put a plate on the strings it will bounce around and
you can't keep it in position. If you put a nail between two
strings, it will slide out. But a wood screw, with its grooves,
will stay. The objects work as mutes, and the sound becomes softer than an ordinary piano, and quite a bit different. I find it as fascinating to prepare a piano as it is to
walk along a beach and pick up shells.

PAGE: Your most famous piece is "4:33", which a lot of people
think is simply the ultimate Cage put-on. Should an ideal
performance of this work take place in a soundproof room?
Are we supposed to hear silence, or is the work's duration
simply a frame into which we can pour the ambiance of
the room and our surroundings?

CAGE: It doesn't matter where you are—whether you are in a
soundproof chamber or outdoors on the streets of Manhattan. There are always sounds to hear. There is no such
thing as silence. That has now been proven scientifically.
There's no way to stop the reception of sound. If you keep
out the sounds coming from outside, then you hear the
sounds coming from inside.

PAGE: But is this a real composition? We traditionally think of composition as the careful ordering of sounds, and here there is no ordering going on.

CAGE: "4:33" was written by means of chance operations. We first performed it in Woodstock, New York, in a theater out in the woods. You could hear the breeze through the trees in the first movement. During the second movement, you could hear drops of rain hitting the roof. In the third movement, people started talking because they realized that the group wasn't going to make any sound, so the sound of people talking became the third movement.

The silent piece actually comes from my study of oriental philosophy—Zen Buddhism in particular. I had noticed that nobody understood what I was doing in my music: I'd write a sad piece and the audience would laugh or I'd write something funny and nobody would even smile. Then I'd go to a concert of works by some other composer, and it wouldn't be clear to me what was happening there. The difficulty of understanding musical language bothered me, so I decided that if music were the language and what we were to do, while listening, was to understand it, then I would just stop writing. I needed to find a more satisfactory reason for composing than to think of it as communication or language or something to be understood. And then a friend of mine, who had been studying Indian music, came up with a new definition for me. She said the purpose of music is to sober and quiet the mind, thus making it susceptible to divine influences. And Lou Harrison found a similar quote from 17th-century England. So this is a proper and tested reason for writing music—not a question of communication or something to be rationally understood, but a question of changing our minds about the fact of being alive.

I became deeply concerned about finding quietude. I knew that, living in the United States in the 20th century, there was no quiet place left, so a quiet mind has to be quiet in a noisy place. I have a friend who owns a lake in northern California and even there he hears airplanes flying overhead. And his refrigerator makes a lot of noise. So we simply can't get away from sound in this century. What we must do is change our minds and hear those sounds with

enjoyment in order to live in a manner that makes us glad
to be alive.

PAGE: But this seems simplistic. You are an expert mycologist—
an expert on gathering mushrooms. And obviously one
mushroom is not as good as another: One will feed you,
the other will kill you. So why are sounds necessarily as
good as each other?

CAGE: What's *not* good is trying to decide which one is good and
which one is bad. The whole concept of value judgment
is a mistake, and if you insist on eating dessert every time
instead of eating your vegetables, then you can just listen
to the sounds that please you. I like to listen to all sounds;
I haven't heard any that I dislike. In fact, I like music less
than other sounds I can think of. I don't like regular beats
in music. I know that my "Third Construction" has a lot
of rhythmic patterns in it, but now I prefer a kind of un-
predictability about where the next sound will come from.

PAGE: I find it very interesting that so many of the minimalist and
jazz composers—most of whom work with a very regular
beat—look up to you as a father figure.

CAGE: Well, I've produced a large body of work. And that work
has different aspects and some of those aspects could lead
a composer in the direction of regular beats. But, for me
and my own experience now, I don't need any music at
all. I have enough to listen to, with just the sounds of the
environment. I listen to the sounds of Sixth Avenue.

PAGE: How do you respond to those listeners and critics who hear
a work like "4:33" and insist that this is not music? Do
you still get angry letters?

CAGE: When you finish a work, it really isn't finished. The person
who finishes a piece is the person who uses it, who listens
to it. Now if somebody is angry with something I've done,
they're unable to finish it, unable to do anything with it.
So, as far as they're concerned, it oughtn't to exist. I received a very angry letter recently, written by a young composer from California who said I had ruined the art of
music for him. I told him that if he had that feeling about
my work, then he shouldn't pay any attention to it. He
should work on what he can do and what he believes it
necessary to do. I've worked that way all my life.

PAGE: One thing that I think is important about your work is that
you've helped us out of the idea of a historical continuum,

which had become a kind of cul-de-sac. Instead of the idea of one forward thrust, one vanguard, as it were, we now accept the idea of a personal music.

CAGE: When I was young, there were only two things that you could do if you wanted to be a serious composer: follow Schoenberg or follow Stravinsky. They didn't even give Bartok a chance! Now, not only because of my work but because of the great changes and interpenetration of cultures, I think you can go in many directions.

PAGE: Where *are* we going? Will music continue to spatter in all different directions?

CAGE: I wouldn't say "spatter." I would say "delta." Instead of a mainstream, we now have a river dividing itself into many streams. There are countless possibilities and they all lead to the ocean. The book Joyce was going to write after *Finnegans Wake* was going to have to do with ocean. Not just tides, not just "here comes everybody" but "here we are, everywhere."

PAGE: One thing Joyce said about *Finnegans Wake* has always intrigued me. He said that it took him more than a decade to write the book, and that he didn't see any reason why it shouldn't take somebody more than a decade to read it.

CAGE: I would hope it would take longer.

(1985)

7 VAN CLIBURN

Fort Worth, Tex.

Van Cliburn—still the most famous pianist in American history although he hasn't played a note professionally for eleven years—stands in the kitchen of his elaborate Tudor mansion outside Fort Worth, presiding over an informal party.

It is 1 A.M. and the night is young. His mother, Rildia Bee Cliburn, frail but alert at 92, sips soup at the head of the long table. Occasionally, unpredictably, she looks up and smiles a warm, satisfied smile, her dark eyes flashing with cricket vitality. Friends—fellow musicians, Fort Worth civic leaders, acquaintances from church—pass through the halls of the gigantic house, dazzled by its riches but doing their best not to sightsee too blatantly.

There seem to be pianos in every room: big, black, glittering grands. And on the pianos are photographs—Cliburn with former President Ronald Reagan, Cliburn with Mikhail Gorbachev, Cliburn at a party and, again and again, Cliburn with his mother. The walls are covered with citations and posters from throughout his career. "Some life, huh?" one of Cliburn's visitors asks as we meditate, solemnly, on the surroundings.

Some life. Cliburn and his mother moved to Texas from New York in 1986 and now live in a mansion once owned by Kay Kimbell, the founder of the city's celebrated art museum. The Cliburns reportedly paid well over a million dollars for the estate—a million in depressed Fort Worth, where one can still pick up a handsome house for less than $50,000. And this is only one of the Cliburn homes: There are places in Shreveport and Tucson, a vast storage space in Santa Barbara. In New York, Cliburn lived at the Salisbury Hotel, where his suite ultimately grew to seventeen rooms. "Van collects antiques and he needs someplace to put them," Susan Tilley, chairman of the Van Cliburn Foundation, explained.

Reputation to the contrary, Cliburn is no recluse in Texas. Indeed,

Fort Worth society clusters to him. The piano competition that bears Cliburn's name brings money, prestige and the world musical elite to Fort Worth, and the city is proud of him—not only for what he has accomplished but for the way he has remained true to Texas.

"He's one of the great men of the world, yet he came back to live here in Fort Worth," one acquaintance told me. "Did you know he still attends Broadway Baptist Church every week?"

Cliburn is gentle, charming and likable. He seems an eccentric but genuinely happy man. He is now fifty-four years old, but the description that comes inevitably to mind is "boyish," and his unaffected use of honorifics like "sir" and "ma'am" only heightens the impression of youth. "Does everybody have everything they want?" he asks in a soft Texas drawl. Assured that they do, he glides on, person to person, group to group, kissing the women on their cheeks, patting the men on their backs, a body in motion, transcendent.

One might say that Cliburn has been in the world but not quite of it for almost a dozen years. He is now almost as well-known for *not* playing as he used to be for playing. He has never completely retired. There have been occasional private performances for friends such as President Reagan, Ferdinand and Imelda Marcos and there was one benefit appearance for the Bob Hope Cultural Center in Palm Springs.

But next week Cliburn will take the stage for his first public concert since 1978. He will perform the first Piano Concerto of Peter Ilyich Tchaikovsky—the work that brought him fame—in a program at the Fredric Mann Center in Philadelphia. And the musical world is expected to turn out en masse to hear this legendary pianist of the past play once again.

Some life, indeed. To find a parallel to Van Cliburn, one must look beyond the rarefied world of classical music and search instead the pantheon of solitary American heroes. Then the analogy comes easily: Cliburn is a musical answer to Charles A. Lindbergh.

Both men were tall, loosely knit, determinedly homespun country boys who were transported, literally overnight, to international celebrity. Both were loners who charted their own courses and then followed them through, unerringly, to fruition. Both men carried an extraordinary amount of geopolitical baggage: Lindbergh was credited with uniting the new world with the old, while Cliburn's victory represented one of the few civil moments in the chronicle of Soviet/American relations during the 1950s.

And, finally, the lives of both men are, necessarily, studies in anti-climax. Lindbergh and Cliburn earned their places in the history books with two or three incredible days in their mid-20s. For all that Lindbergh accomplished, for good and ill, after The Spirit of St. Louis made its way across the Atlantic Ocean, everything pales when set beside those thirty-three airborne hours.

And, whatever Cliburn may do with the rest of his life, one suspects he will always be remembered as the Texas boy who found his way to Moscow at the height of the Cold War and brought home the gold— first prize in the Soviet Union's most prestigious musical competition, the Tchaikovsky.

Cliburn, who was then twenty-three years old, had already won several important American competitions, among them the Kosciusko Foundation's Chopin Prize and the Leventritt Award. But his career had begun to falter. By early 1958, he was living back home in Kilgore, Texas, tending house for his parents (Cliburn's father was a purchasing agent for a local oil company). The decision to go to Moscow had the desperate quality of a last chance.

His success was extraordinary, unprecedented and reported on the front pages of newspapers throughout the world. At a time when there seemed no common ground between the United States and the Soviet Union, Van Cliburn provided one. He was mobbed in Moscow with a fervor of startling intensity. Women wept and fainted at his concerts; Khrushchev himself embraced the young American.

Time magazine put him on its cover with a banner that read "The Texan Who Conquered Russia." Fans ripped off the door of his limousine during a visit to Philadelphia. RCA Victor signed him to an exclusive contract and his first recording—the Tchaikovsky Piano Concerto No. 1, of course—quickly became the best selling classical disc of all time, a position it retained until "Switched-On Bach" was released a decade later. By the time he was twenty-four, he was the subject of a biography, written by the late critic, composer and pianist Abram Chasins. It was called *The Van Cliburn Legend*. Few young pianists have ever had so many expectations to live up to.

Not surprisingly, Cliburn apparently found it an impossible task. "From the mid-1960s, it seemed that he could not cope with the loss of freshness," Michael Steinberg has written in the *New Grove Dictionary of Music and Musicians*. "His repertory was restricted; his playing, always guided primarily by intuition, took on affectations and the sound itself became harsher."

In 1978, Cliburn played his last concert and settled down to live quietly with his mother in New York. Nobody knew if he would ever play again.

Like many aerialists, Cliburn seems a little uncomfortable with life on the ground. Sitting him down for a talk—a *serious* talk, beyond flutter and pleasantries—is all but impossible. ("You don't interview Van, you experience him," one friend told me.) Cliburn is always terribly polite and never quite says no, he won't give an interview. He would rather change the subject or give a diplomatic, courteous non-answer and breeze on, nothing revealed, no positions taken. Even an innocent social question about whether he misses New York is handily depersonalized: "Well, New York is one of the great cities of the world." And the subject is closed.

But late that evening—or rather, early the next morning—Cliburn reluctantly sat for an interview. The party safely in swing, he closed off the music room (two pianos instead of one) and settled into a comfortable, overstuffed couch to discuss his long absence from the concert stage.

"I can remember when I was about eighteen and some friends of mine were going to Europe for the summer and they asked me to come," he said. "And I thought I'd get there to play sometime anyway, so I didn't go. You know how when you're young, you do all this prognosticating which later turns out to be right or wrong. And I told my friends that I would work very hard for the first part of my life, and then I'll take an intermission and then work very hard the last part of my life.

"Well, anyway, I was very busy for a long time," he continued. "And then, in January 1974, my father died. And then less than two months later, Sol Hurok died. I had been supposed to see Mr. Hurok in New York in January, but I canceled because of my father's death. So he said, 'Oh, when you come next month we will talk.' But I missed having that last lunch with him because he was dead by the time I came back to New York.

"And I realized that I was missing lots of little lunches and dinners with my friends, and that too many of those friends seemed to be departing. And I decided I wanted a little bit of time to enjoy my friends because things don't last forever. I want some personal memories too, as well as the memories of meeting wonderful people in the concert halls and interesting times traveling. So I planned in 1974 to take some time off but never announced it. I just stopped accepting engagements. Because my wonderful memories of concertizing are so vivid, I feel like I've never left. It's like the intermission was a little longer than I thought, but it's all still part of the same concert.

"You know, I love to be able to go to concerts and hear my friends perform," he said. "That really pleases me, and I really wasn't able to do that for—oh, the longest while. I made my orchestral debut with the

Houston Symphony when I was twelve. And then I was playing lots of recitals throughout what we call the Arklatexas area. And then I made my debut with the New York Philharmonic when I was nineteen. So from childhood on, it was a lot of work. I've enjoyed my time off immensely. It's been a grand time, and I've had an awful lot of fun."

One cannot escape the suspicion that the "time off" may ultimately prove to be a permanent state of affairs. For Cliburn is vague about every aspect of his return to the stage. "I'm not going to be playing a million concerts again, I can promise you that," is about as definite a statement as he will make.

He says that the reason he is playing in Philadelphia is largely sentimental. "It's a tribute to two very good friends of mine who have died recently—Eugene Ormandy and [patron of the arts] Fred Mann."

When Cliburn is pressed about pieces of music he wants to learn or to perform, nothing comes to mind (he is playing his signature piece, the Tchaikovsky, in Philadelphia). Recordings? "Well, yes, I plan to make some records, but I haven't really gotten into the planning stage yet." A solo recital? "I have some lovely offers, and I may well just take one of them up." Other pianists? "I thought the four that we had here last night were all wonderful." If Cliburn feels an urgent need to make music, he keeps it to himself.

I was reminded of one of Aaron Copland's last interviews, 10 years after he had written his last major piece. "I'm amazed I don't miss composing more than I do," Copland said. "You'd think if you had spent 50 years at it you'd have the feeling that something was missing and I really don't. I must have expressed myself sufficiently." And so, perhaps, has Cliburn.

This is not to suggest that Cliburn cannot perform, only that, from the point of view of a prodigy who was driven hard from the beginning of childhood, it may well be liberating *not* to. One suspects that he will continue to play occasional benefits, visit the White House now and then; there is talk of a return visit to the Soviet Union—all circumstances where the principal interest is extramusical.

The rest of the time, there will be the loving friends, the sparkling parties, the fulfilling church life, the adulation of his native state, the challenge of the Van Cliburn Competition and the legacy of some fine recordings (now in the process of reissue on RCA Victor compact discs).

Whatever may happen in Philadelphia next week or in the years to come, Van Cliburn knows that his spot in history is assured.

"I remember calling my mother from Russia and telling her I'd won the competition," he said with a grin. "I had no idea that the story had become so big. So I asked her if she'd told Mrs. So-and-So across town that I'd won. And she said yes, she knew all about it. And I asked her if she'd told Mr. So-and-So in the next town over that I'd won. And she said, yes, he knew about it, too. And I felt pretty good, because it was all very well to be known in your town, but what *really* mattered was if your reputation had spread to the next town. Then you'd really made it."

He smiled and shook his head, and there was a note of awe in his voice. "I was mighty proud that night," he said.

Newsday (1989)

8 CHRISTOPH VON DOHNANYI

Cleveland is a city of contrasts—of grandeur and deprivation, purse-proud burghers and working class heroes, glittering skyscrapers and austere clapboard houses, poetry and grime. It was here, in what the local chamber of commerce used to call "America's Sixth City" before the population diminished, that John D. Rockefeller made his first million, that the young Hart Crane scratched down kaleidoscopic verse in the turret of his family home, that an obscure disk jockey named Alan Freed introduced a subversive new form of music called "rock and roll" on his radio show.

Paradoxically or otherwise, it is also in Cleveland that one finds what may now be the finest orchestra in the United States.

This week, the Cleveland Orchestra, under the leadership of its music director, Christoph von Dohnanyi, will present four concerts in and around the New York area—at Tilles Center on Long Island and Carnegie Hall.

Dohnanyi's podium style is neither effusive *à la* Leonard Bernstein—whose method approaches choreography—nor is it as tautly economical in the tradition of George Szell and Fritz Reiner. Dohnanyi's interpretations tend to be calculated, disciplined, transparent, unsentimental but deeply committed. He is a reserved but demonstrative conductor; players and audience have no difficulty following his beat. And the orchestra plays with reflexive power and clarity for him.

Formal, urbane, meticulously groomed, with iron-gray hair, Dohnanyi looks rather like Tom Brokaw transformed into a Middle-European count. He speaks English better than he thinks he does, albeit with a marked accent. "I *am* European—in my upbringing, in my perception of the world, in my soul" he said one recent morning, over tea at the Union Club in downtown Cleveland, a grand, sooty-faced Victorian where one feels naked without a tie.

"I spent the first 50 years of my life in Europe; hence I am a European. That can never change. But I am so taken by life in the United States; I enjoy the people so much. Americans don't make things complicated when they are not. They are *positive*, you know; even if there is a depression, they know they will come out of it eventually.

They think of solutions; they don't complain all the time. That is what I love about this country. Europe is afraid, reluctant, haunted by the glory and tragedy of its past."

Dohnanyi's own history embodies both Europe's glory and its tragedy. Born in Berlin in 1929, he grew up in a cultivated and liberal household, and began his musical studies at the age of five. His grandfather was the Hungarian pianist and composer Ernst von Dohnanyi, whose son married a German and moved to Berlin. Christoph's father, Hans von Dohnanyi, was eventually executed for anti-Nazi activity—an attempt on Hitler's life, in fact, one that almost succeeded. His uncle, the theologian Dietrich Bohnhoeffer, was also murdered in a German prison camp.

After World War II, Dohnanyi enrolled in the University of Munich, with the intention of becoming a lawyer. But he abandoned his law studies when he was awarded the Richard Strauss Prize for Composition and Conducting in 1948. There followed a period of intensive study— with Bernstein at Tanglewood, and with his grandfather, who had come to the United States to be a professor of music at Florida State University.

Dohnanyi's first professional engagement was with the Frankfurt Opera, where he served as a coach and conductor under Georg Solti; eventually he became the company's artistic and musical director. From 1978 to 1984, he served as the artistic director and principal conductor of the Hamburg State Opera.

Despite these accomplishments, Dohnanyi was the darkest of dark horses when he was appointed music director in Cleveland. For, although he had at this point built up an impressive reputation with his work in the opera house, he was not particularly well known as a symphonic conductor.

He was also considered something of a Modernist, with a strong commitment to 20th-century music, which rarely goes over well at the box office or in the boardrooms. Moreover, some felt that such a prestigious job should go to an American. (Critics seem to care quite a bit more about this particular qualification than symphony boards: Not one of our so-called "Big Five" orchestras currently has American leadership.)

Along with the symphonies from New York, Boston, Chicago and Philadelphia, the Cleveland Orchestra has been ranked among the "Big

Five" for some forty years. It is an antiquated and fairly arbitrary list—some listeners would argue that St. Louis, Pittsburgh or Los Angeles deserves a place in the roster—and, God knows, all of these groups have bad nights. But they are also, at their best, genuine *ensembles:* congregations of men and women who transcend individual ostentation for a greater glory.

Unlike the Boston and New York symphonies, which have been good but fairly generic in recent years, the Philadelphia and Chicago orchestras really do possess distinctive *sounds.* In Chicago, all is aggression, flash, intensity, virtuosity. It is probably reductive to suggest that the orchestra represents the big brassy utterance of a big brassy town—but there it is. Meanwhile, the Philadelphia Orchestra is what the critic and composer Virgil Thomson once referred to as a "string combo" and has been famous for the lush, velvet smoothness of its sound since the days of Leopold Stokowski.

In its present estate, the Cleveland Orchestra combines the best qualities of the Chicago and Philadelphia orchestras. It can summon the brilliant sonorities associated with Chicago, but there is nothing vulgar about the endeavor; to put it bluntly, the Clevelanders never use music as a pretext to make a dazzling, deafening noise, as their neighbors to the West are suspected of doing. And the Philadelphia sound is sometimes a bit *too* creamy and caloric: Cleveland adds some welcome muscle tone to all that elegance.

The Cleveland Orchestra, founded in 1915, is by far the youngest of the "Big Five." The orchestra's early conductors included Nikolai Sokoloff, Artur Rodzinski and Erich Leinsdorf. But it was only during the tenure of the exacting, autocratic George Szell, music director from 1946 until his death in 1970, that Cleveland attained world prominence.

Szell was, by all accounts, a difficult man (Rudolf Bing, the long-time general manager of the Metropolitan Opera, when informed that "George Szell was his own worst enemy," snapped back "Not while I'm alive") but he took a provincial ensemble and built it a great orchestra. He enlarged membership to 107 players, extended the symphony season to a full year, and left a legacy of fine recordings, many of which are still in print.

In 1972, two years after Szell's death, Lorin Maazel was appointed music director. He held the position for 10 years, during which time, in the opinion of many listeners, the orchestra's standards precipitously declined. In 1982, Maazel left Cleveland; the usual amenities were observed between conductor and management in their respective statements, but it was clearly time for a change. And so on to Dohnanyi, and to what may well be the Cleveland Orchestra's finest hour.

Dohnanyi, who assumed the directorship in 1984, quickly won a following—with the players (by firm but thoughtful leadership), with the board (through an increase in orchestra subscriptions and recording contracts), with the critics (through creative programming and assured, majestic performances) and with the city of Cleveland (by purchasing a house in the area and taking a relatively active role in civic affairs).

Currently, Dohnanyi spends about nineteen weeks a year at his home in Shaker Heights, an affluent suburb to the southeast of the city, where he lives with his wife, the soprano Anja Silja, and their three children.

"This is my artistic marriage," Dohnanyi said of his post in Cleveland. "I would never take on another orchestra, no more than I would another wife. Indeed, I do very little guest conducting. I feel that I am taking care of the Cleveland Orchestra, even when I'm not here."

Dohnanyi has also maintained a home outside Hamburg but, while most of his European performances now take place in opera houses, he is unlikely to conduct in his native city.

"I had tremendous difficulty in Hamburg," he said recently. "The thing I hate is when opera orchestras say: 'Today you'll have these players, tomorrow you'll have those, and we don't even know who is playing for the third rehearsal.' I had to fight it. For one thing, intonation suffers. A first oboist who does not know his partner is always thinking: Is he high on that G or is he flat?

"In Germany you think you are being invited for a nice meal and suddenly you discover you have to clean the dishes first. Before anything else, you have to install a C major chord in tune!

"One thing I love about the United States is that when you ask for a C major chord, you get a C major chord," Dohnanyi continued. "There is an incredibly high level of proficiency in American orchestras—much higher than in Europe. There's never any problem with intonation, so you can go on from there.

"Now Szell is responsible for this level of discipline," he explained. "And then Maazel turned the Cleveland into a wonderful reading orchestra: He was restless and wanted to do lots of new things without much rehearsing—just play, play, play. What I want to bring to Cleveland, which maybe it didn't have before, is a sense of breathing, *musical* breathing, like a singer. I sometimes think that Szell took one breath before the performance and then another after it. There was always this electric, inner tension in his performances, but it was occasionally dangerous to the music making. Szell was a master, you know, but I think he may not have been so fond of singing."

Dohnanyi relaxed and sipped some tea. "I *am* fond of singing. Once you've conducted a lot of opera, you start to hate bar lines. Singers never think about them and I try not to think of them when I conduct. There are some conductors who play measure by measure, metronomically, and they are very reluctant to cross bar lines. But I want to take the music and let it flow freely, taking advantage of the wonderful facilities of a modern orchestra. I don't even mind a little insecurity about the rhythm if it makes for additional flexibility."

This calls to mind the apocryphal story about the German conductor Wilhelm Furtwängler, whose downbeat was notoriously irregular. When one of his violinists was asked how they followed Furtwängler, he said "We wait thirty seconds after he mounts the podium and then start to play."

Dohnanyi laughed. "I'm sure it isn't true, but the story does capture something of Furtwängler. He was a great hero of mine—nobody has ever come close to his Bruckner, nobody!—and I heard him many times. Of course he had technical shortcomings. But art is beyond technique.

"What interests me is the event, the concert, the recording, whatever happens in the performance. I don't care how the performance is achieved. And, despite his technical problems, most of Furtwängler's performances turned out splendidly, because the orchestra knew him, knew what to expect. This is another reason why I believe in commitment to one orchestra. Conductor and players learn to understand one another, quirks and all."

Which brings him to the subject of the newly renovated Carnegie Hall.

"I'm convinced that one of the reasons that there has been so much debate about the acoustics of Carnegie Hall is because there is no house orchestra," he said. "Nobody really knows what the hall sounds like because somebody different is playing there every night. It is a guest house now, since the Philharmonic moved out—and I am convinced that one of the reasons for the Philharmonic's bad press is the fact that people only hear it play at Lincoln Center.

"My own opinion is that the new, refurbished Carnegie is not so terribly different from the old, but if there are changes to be made, I wish the management would go ahead and make them. I don't believe in those acoustic panels on the stage"—a measure taken by the Carnegie Hall administration at the beginning of the 1988–89 season, in an attempt to improve the sound—"they look ugly and I'm fed up with them.

"You know, my only complaint about Carnegie Hall is an old one— a problem that was there before the renovation and afterwards—and that is the subway running underneath. It disturbs me tremendously. I cannot play my Cleveland pianissimo, of which I am very proud, in New York. All you would hear is a rumble."

When asked about the conductors whom he esteems, Dohnanyi instantly volunteers the names of Furtwängler, Hans Rosbaud ("People always thought he was just for modern music, but he was extraordinary"), Toscanini ("for Italian music") and Herbert von Karajan. "I am only sorry that many of Karajan's later recordings are so sloppy; there are just mistakes all through them.

"And I admire Leonard Bernstein, in some pieces," he continued. "At the very least, I'm always surprised by Bernstein. He is at his best in material that you might not expect him to find sympathetic—in Haydn, in Schumann, for example, just wonderful. But I am absolutely opposed to his Mahler. He does to Mahler what Mahler did to Beethoven: overdoes it, changes it around, blows it all out of proportion. And his new recordings are simply too slow. I don't think they work at all. But he is a real genius, a tremendously musical man. Even when I think what he is doing is a little. . . . "—he pauses, then smiles politely, the right word at hand—"unusual."

Dohnanyi has an affinity for contemporary music of all kinds. Last year, for example, the Cleveland Orchestra commissioned a new work, "The Light" from Philip Glass, whose bright, tonal, aggressively repetitive patterns have little to do with the angst-ridden chromaticism of so much postwar European music. Dohnanyi and the orchestra played "The Light" in Cleveland and New York; at Carnegie, they were rewarded with the mixture of boos and bravos that seem to follow Glass around.

"I think that Dohnanyi was a little bit perplexed by my piece," Glass said recently. "This really isn't his sort of new music at all. But he was a complete professional—courteous, helpful and insightful— and I was knocked out by his performance. You heard it? Just incredible."

Leonard Slatkin, the music director of the St. Louis Symphony, is also an admirer of the Cleveland Orchestra. "It's probably the most conscientious orchestra around when it comes to matters of detail," he said. "The players really want to get things as clean and well organized as they can. Whatever the conductor wants, they provide."

Ronald Bishop, the principal tuba player in the Cleveland Orchestra since 1968, when George Szell tapped him from the San Francisco Symphony, gives Dohnanyi high marks. "He reminds me of Szell, in that balance is extremely important to him. He always wants to hear all the voices. He is also very conscientious in honoring the composer's wishes." Dohnanyi's rehearsal style, according to Bishop, is "cordial, gentlemanly and professional."

Jeffrey Khaner, the principal flutist, agrees. "I think Dohnanyi's management of rehearsals is one of his strongest points. He is very orderly, has a good sense of what he wants to accomplish and then goes about it in the most businesslike way. The morale of the orchestra is very good. We are delighted to make so many recordings. And I, personally, am very happy about all the unusual contemporary music that we do."

Dohnanyi's approach to new music has been described as "intellectual"; it is, at the very least, analytical. "What I do is destroy first," he told an interviewer. "I question everything. Schopenhauer once said: 'Wisdom is what remains when you have forgotten all you have learned.'

"I am very selfish, really" he explained. "I do not want to specialize. I want to conduct *everything*. I don't want to specialize in contemporary music but I could not live with myself if I did not understand it. If I read Marx, I must learn about the reflection of Marx in our own era. And the music of our day reflects the music of the past."

It's opening night of the orchestra season at Severance Hall, some five miles to the east of Public Square, the spiritual center of Cleveland. One rides out Euclid Avenue through a panorama of urban life—refurbished office buildings, dilapidated Charles Addams-style mansions, hospital complexes and piles of rubble. Then, suddenly, just off the campus of Case Western Reserve University, there is Severance Hall, an extraordinary art deco *cum* movie palace creation, planned in the last days of the 20s, and finished shortly after the crash, by which time it was already a magnificent relic.

Cleveland society was out in style—students and senior citizens, tuxedos and blue jeans. The orchestra members were dressed in formal black, of course, as was Dohnanyi, who walked onto the podium a few minutes after eight, greeted the audience with a reserved smile and a deep bow, then turned and lifted his baton.

What followed was decidedly out of the ordinary for such a festive occasion. The opening piece of the season was not some popular curtainraiser by Brahms or Elgar but rather a work by the Hungarian composer György Ligeti, who is best known in this country for his buzzing, swirling "Requiem," used effectively in the film *2001: A Space Odyssey.* "Atmospheres," which dates from 1961, shimmers from silence, clusters of sound meeting and melding through time and space. One is put in mind of a ballet of clouds, captured in time-lapse photography. Strings give way to brass, then winds, in an elaborate musical process that is as orderly as it is colorful.

After some ten minutes, "Atmospheres" died away. And then, in a segue worthy of the most inspired radio programmer, Dohnanyi led directly, without a rest, into the hushed, prayerful prelude to Wagner's *Lohengrin*. It seemed to emanate from the afterglow of the Ligeti and, by contrasting the two works in this way, letting them enrich and cross-fertilize one another, Dohnanyi heightened the potency of both.

"This was the first time I ever did it that way," Dohnanyi said the next day. "And it worked. After the absence of any tonality in the Ligeti, one hears music differently when tonality is suddenly, almost imperceptibly, added. It's somehow a different experience—something you don't take for granted any more. If I had programmed the Wagner first, you would have had the audience relaxing through this nice piece in A major, and then waiting patiently through the Ligeti and its dissonances.

"It may take time," Dohnanyi continued with quiet determination, "but I do not want to underrate my public. Even if people are reluctant to listen to new music at first, if you challenge them, and present the music intelligently, they *will* respond.

"I want to be challenged, too," he said. "This is the reason I rarely play my grandfather's music. His best work is his chamber music, and that is very beautiful. But the orchestral pieces . . . I don't like music that I understand. Grandfather's music was beautiful but I understand it. And I really love music that remains a challenge, and two bars of Mozart, Bartok, Webern, Bruckner remain a challenge for a lifetime.

"You know, ultimately the only important artists are the creative ones. I consider myself—all of my conducting colleagues—necessary only as a keyboard is necessary for a piano. We should do our best, and love and honor our profession, but the ones who count are the composers, the painters, the writers, the architects. Those are the people I honor."

Newsday (1989)

9 MATTHEW EPSTEIN

Liverpool, England

As far back as 1909 that devoted Anglophile William Dean Howells, in the chapter title of one of several books he published about England, was only able to work up "A Modest Liking for Liverpool." German bombing raids, the decline of the shipping industry and what has become a more or less permanent recession have not improved matters. In the same way that Detroit, rightly or wrongly, epitomizes America's blighted cities, gray, gloomy Liverpool has become the standard by which Britons measure urban decay.

And yet, the 2,500-seat Empire Theatre, in the town center next to vast Lime Street Station, is filled to capacity this cold December evening, and the mood, offstage and on, is one of giddy exhilaration. Although the poster cases outside the hall are mostly given over to male torsos and Day-Glo lettering ("Chippendales is Coming!" "Mr. T. as Aladdin!"), the attraction tonight is rather more rarefied: Welsh National Opera, a caravan of buses and trucks has come up from Cardiff to present Johann Strauss' *Die Fledermaus.*

Matthew Epstein, 44, famous in New York opera circles for two decades, a vice president of Columbia Artists since his early 20s, artistic adviser to, among others, the major opera companies of Chicago, Santa Fe and San Francisco and since August the general director of Welsh National Opera—the first American within memory to attain such an exalted position with a European company—sits in the balcony, watching nervously, whispering a running commentary on the performance.

"The narrator is all wrong, *all wrong,* for this, and she should be wearing a pants suit instead of evening dress... The spotlight is too much. Out, damned spot!... The conductor's doing a very nice job; yes, sounds a little like Rossini, doesn't it?... Now this soprano has world-class potential. Listen to that voice! She isn't yet quite sure what to do with it, but she'll learn. Listen!"

It is, unquestionably, a good show, made all the more remarkable by the fact that WNO is a traveling band, and must adapt accordingly. One night the targeted audience will be thousands of people in Liverpool, the next night the same opera may play to fewer than 700 at a tiny theater in Swansea. WNO also regularly services Bristol, Oxford, Southampton and Birmingham and pays the occasional visit to London; in 1989, it even came to the Brooklyn Academy of Music, with its Royal Patron, Her Highness the Princess of Wales (better known stateside as "Princess Di") and an acclaimed staging of Verdi's *Falstaff.*

When the opera ends, Epstein sounds the first applause. It is quickly seconded, and we are suddenly in the midst of a very proper, British pandemonium. Ten minutes later, as the crowd dissipates and downtown Liverpool takes on its customary evening quietude, Epstein walks out in front of the stage to meet with those listeners who have accepted an open invitation to stay and talk opera.

"So what did you think?" Epstein asks eagerly as he greets the 100 or so people who remain. "I want your questions, your comments, your advice. After all, I'm the new boy here."

An obvious statement, this last. With his casual attire—slacks, sports jacket, sweater ("I hate tuxedos and won't wear them"), determined informality of manner and long, crinkled hair pulled back into a ponytail, Epstein stands out in a crowd, particularly a formal, English, first-night-at-the-opera crowd. And yet, this is no upstart escapee from standing room but *the boss himself*—the man in complete artistic control of the company many aficionados would rank the liveliest in Great Britain. When the final question has been answered and the Empire Theatre is deserted but for the stagehands packing up the production for a 3 A.M. departure, Epstein walks through the station on his way back to the hotel.

"Everybody has been terribly supportive, from the board of directors on down, but I have no doubt that some people over here don't know quite what to make of me," Epstein explains, tired but good-humored at the close of what has turned into one more 18-hour day. "After all, I'm an American. I'm a New Yorker. I'm Jewish. I'm fat. I'm gay. And I have AIDS."

Epstein believes he has been infected since the early 1980s; when a test for the HIV virus became available in 1986, he immediately took it and was not surprised by the results. He now begins every morning with what he calls his "cocktail," two big tablets of a recently FDA-approved drug known as DDI, pulverized, stirred into a glass of water and swallowed with a grimace, a process repeated in the late afternoon. On every second day, he must take a Bactrim tablet to prevent HIV-related pneumonia. He has mastered the code language people with AIDS necessarily develop. He monitors his T-cells, keeps track of his platelet count, visits doctors in New York and London, does his best to stay healthy. With the exception of chronic fatigue and consistent irregularities in his blood count, Epstein is asymptomatic. But, as a reminder of the odds facing him, he carries a list of friends who have died—more than 75 of them to date—wherever he goes.

Two weeks after Liverpool, Epstein was back home in Manhattan; having added another several thousand miles to what is already an extraordinary frequent-flyer account (he crosses the Atlantic about 20 times a year and is present at most important opera productions throughout the United States and Europe). During his two-week stay, Epstein would:

• Prepare for the presentation and national telecast of a Rossini bicentennial celebration he is producing at Lincoln Center on Feb. 29, featuring Marilyn Horne, June Anderson, Rockwell Blake, Samuel Ramey, Thomas Hampson, Chris Merritt, Frederica von Stade and others with the Orchestra of St. Luke's conducted by Roger Norrington.

• Help plan the third "Music For Life" concert in 1993, a gala benefit for Gay Men's Health Crisis, the oldest and largest AIDS service organization in the country, that is expected to raise more than a million dollars in the course of the evening.

• Handle several complicated business arrangements for the soprano Kathleen Battle.

• Coach, lecture, nurture and fortify several singers, famous and soon-to-be-famous.

• Meet with representatives from the Chicago, Santa Fe and San Francisco opera companies to discuss upcoming season.

• Attend the premiere and post-premiere party for *The Ghosts of Versailles* at the Met.

• Serve as the host of his own New Year's Eve bash that started at 11:30 in 1991 and ended well past the first dawn of 1992.

And more.

"I thrive on stress, on pressure," Epstein said over a Sunday brunch in the Ansonia Hotel suite on the Upper West Side he has occupied since 1973. "I can't stand it when things are too calm. But there is a limit, and I've learned that limit. I've finally learned to say no, I *can't* do anything next Friday, because I'm taking the day off; I *can't* come to that meeting, because I need to stay in bed another hour. That kind of thing. If I'm really pushed, I'll just get angry and say damn it, leave me alone, I'm *sick!*"

Still, it is obvious, to anybody who spends even an hour with Epstein, that this is a man who loves his work—somebody who combines an inexhaustible, almost childlike, appetite for music theater with an encyclopedic command of its intricacies.

Epstein grew up in New York and Long Island, and began frequenting the Metropolitan Opera before he was in high school ("Standing room was $1.50 upstairs, $2.50 in the parquet," he reflects with a mix of nostalgia and wonder). He heard Maria Callas in her last New York *Tosca*, saw Fonteyn and Nureyev dance *Swan Lake*, attended Vladimir Horowitz' famous 1965 return concert at Carnegie Hall and countless Philharmonic concerts under Leonard Bernstein. "It was a great time to be in New York," he says now. "I was a lucky kid."

He was also, by his own admission, an aggressive kid. While still a teenager, he began to visit singers backstage after performances, astonishing them with frank assessments of what had gone right—and, more delicately, what had gone wrong. "*Of course* some of them took it badly—some of them took it *very* badly," he recalls with a mischievous grin. "But the real artists were grateful that somebody would tell them the truth. There are a lot of sycophants out there."

A legend of sorts formed around the intense, opinionated young man who heard—and said—what the seasoned tastemakers of the opera world could not, and soon Matthew Epstein, the inspired, impassioned amateur, turned pro. As adviser, as agent and, finally, as impresario, he fought for the American singer.

Marilyn Horne was an early friend and admirer. Epstein signed Frederica von Stade to her first contract while she was still a student at the Mannes College of Music. He worked closely with Evelyn Lear, Thomas Stewart, Samuel Ramey, Kathleen Battle, Catherine Malfitano, Neil Shicoff and dozens of others at crucial points in their careers. And by 1973 he was running his own division of Columbia Artists Management Inc., the most powerful agency in the music business.

Epstein is one of the unifying threads in Ethan Mordden's diverting 1985 study of opera mania, *Demented*. After noting that the 19th-century impresario James Henry Mapleson once described himself as a "student, critic, violinist, vocalist and composer, concert director and musical agent," Mordden adds: "There is no place for a Mapleson today; now, only the artists freelance. His metier, moreover, has broken into its constituents, yielding one profession for the drafting of singers' schedules, another for the guiding of singers' careers and a third for the artistic and/or regulating steerage of the resident companies.

"Only one person has thought it useful or amusing to build the three jobs together again, the American agent-plus-manager-cum-impresario Matthew Epstein," Mordden continues, "Applying an almost mystical knowledge of opera from core repertory to potential rediscoveries, Epstein was, as agent, the most demanding, when guide, the most insightful; when impresario, the most comprehensive."

Somewhere within the recesses of the improbable Brittania Adelphi Hotel, a great, bleak battleship of a building adjacent to the opera house in Liverpool, Claude Robin Pelletier, a young tenor from Quebec and a member of Welsh National Opera, makes an appointment to talk with Epstein about his career and receives the lesson of his life.

"Look, here are the best managers and public relations people in the States, the people you must see," Epstein snaps out bang-bang-bang—phone numbers, addresses, artistic predilections—faster than the flabbergasted Pelletier can write them down. "And use my name. I never allow anybody to do that, but *you* can. I think you could do well in New York. But don't forget that there's a lot to be done in Europe. Eastern Europe, *East Germany!* It's going to explode! Once the economy calms down, they're going to need people to fill all those theaters, and they'll no longer have to pick and choose within the Eastern Bloc. It's a whole new ball game.

"What have you been doing since I last saw you?"

Among other things, Pelletier has signed on to stand by, or "cover," for Neil Shicoff in a Toronto *Carmen.*

"Good, good, but we have to get you out of Canada. Face it, you've *done* Canada."

There is a *Romeo et Juliette* coming up.

"Romeo, eh? Well, you're the type for the part, and you're excellent onstage, but it's a *big* role. Remember, Corelli sang it. Jean DeReszke sang it, and he was also a famous Tristan. But do it if you want. Every year, sing something big, but then go right back to your repertory—the standard, central tenor repertory. If a house wants you for something different, tell them, yes, I'd love to come to your theater, but I must also have Tamino [in *Magic Flute*] or Alfredo [in *Traviata*]. That's how you build a repertory."

Deep breath.

"What do you sing at auditions?"

Rigoletto.

"The Duke's big aria? Good. What else?"

Tamino.

"Yes...but *auf Deutsch.* What is your best French aria? Faust? Okay, but *with* the high C. Why don't you try *Mignon* and *Fille du Regiment*

... not the first aria but the slow one, that's very beautiful. And Lensky's aria—in the original *Russian*. Yes, you can do it. Just give it a try.

"Remember, if they ask you to sing the wrong role for your voice, just say no; you've heard that phrase. And if your coach tells you nothing but bravo, bravo all the time, go get another coach. You need to work on the art with somebody who will take you seriously.

"We'll talk in the spring," Epstein says as he shows the overwhelmed Pelletier out with a firm pat on the back. "Why don't you go to Germany? I think that's your best move right now.

"Be smart," he calls into the corridor. "I *challenge* you."

The door closes, the smile wilts and Epstein suddenly pales. "God, I feel *terrible*." He falls into an overstuffed chair and shuts his eyes; for a moment, the crucial trajectory that keeps him going, hurtles him ever onwards, visibly falters. "But... it's show time in an hour." He rises to his feet, a new store of energy coursing into his body, ready for the long night ahead. "Shall we go?"

It is estimated that one in five Americans knows somebody infected with the HIV virus. And yet, for many—especially those who live outside the designated "risk groups" and/or a major metropolitan area—the disease still seems as remote as the plagues that ravaged 14th-century Europe. Anybody involved with the arts, however peripherally, knows the real story—knows one of the many composers, conductors, dancers, musicians, actors who have died young, on the threshold of valuable careers; knows, perhaps, that the New York City Opera alone has lost more than 50 people to AIDS since 1981.

Epstein is aggressively candid about his medical condition. "This is a war, and the more people who come out and join the troops the better it'll be," he said. "I believe that being homosexual and closeted today is a little like being a German Jew in 1939 and changing your name and hiding your background. To me, it's immoral, *deeply* immoral, at this particular moment in time."

And so the Right Honorable Lord Davies of Llandinam, chairman of WNO's board of directors, was duly informed of Epstein's status as their negotiations commenced. "We put some thought into it," Lord Davies said in a recent interview. "We got some good medical advice, learned that there was at least the probability that he would be able to fulfill his contract, and then made a unanimous decision to proceed with Matthew, because, of all the people we met with, he was the one we really wanted."

Epstein's initial contract is for five years, with an option to renew. "I plan to keep going as long as I can," Epstein says. "When I first became ill, I went to see a very good therapist, and she told me that the only way to keep alive was to remain deeply involved with something you love. If you give up your life force, you give up your life. That force varies from person to person—it might be writing, research, a profession, another human being. With me, it's music.

"And then I have a lot of support from my friends. We make an effort to stay in touch. We call. We see one another. I visit a city where I know I have some friends, and we prepare a small dinner party. We don't go to bars much anymore—I never really spent much time in bars, anyway. And when I went to the baths, I usually felt like an observer— you know, *this* is interesting, look at all those pretty people. Either that or I'd feel a wistful sadness, a sense that this couldn't last. We took it rather far. I was never very promiscuous, though. I'm not making any moral judgment; I just *wasn't*. During the wildest years of the sexual revolution, I was much too busy building a career to stay out partying until three in the morning.

"Still, I'm sick. And it is likely that I contracted this disease from encounters I had—several of them, probably—with people who are now gone. And what's happened in the last ten years is that the world has become afraid of sex again. You can see it in the faces on the street, men and women. And that's a crime, because sex is a joyful thing. It's like music. It's like food."

A visitor, groping awkwardly to offer some encouragement, points out that people with AIDS are living longer than they once did. A number of individuals have now been infected with HIV for more than a decade; perhaps research will ultimately enable Epstein's health to be maintained indefinitely.

"One would like to think so, but I don't believe it," he answers gently. "There'll be a vaccine some day, but I don't see any cure for a while. And so I believe I'm a walking time bomb. I could become violently ill at any moment. You see, what's happening now is that people aren't just coming down with Kaposi's sarcoma and pneumonia anymore— the traditional manifestations of AIDS—but with weird bone diseases, brain disorders, all sorts of mysterious stuff. I had one friend who was in the middle of a phone conversation when he suddenly lost the power of speech. And that was it, you know. He was able to walk, able to move around. He just couldn't speak any longer, because something had happened to his brain."

Twilight on the Upper West Side, a few days before Christmas. Epstein plays back his phone messages as bells from a nearby church toll serene counterpoint. A woman's voice wants to know what "Kathy" will be wearing in an upcoming telecast (this being Matthew Epstein's answering machine, "Kathy" means Battle). A man with a thick accent *must* know—*immediately*—about some contracts. There are calls from Berlin and Cardiff, and Teresa Stratas, a friend and neighbor at the Ansonia, has slipped a bread-and-butter note under the door.

Epstein smiles as he reads the familiar, elegant script and then sets the card on his grand piano. "You know, I'm genuinely happy in a lot of ways. I've wanted all my life to run an opera company, and now I've got one of the best in the world to work with. And I have *tons* of ideas— singers, directors, conductors I want to employ, commissions I want to make, operas I want to stage.

"Still, whenever I'm planning a production—and they have to be arranged two or three years in advance—I can't help wondering if I'll be around to see it happen."

He shrugs and walks into the kitchen to grind another dose of DDI.

Newsday (1992)

10 GERALDINE FARRAR

Much of my tenth year was spent lost in the antique sounds of old opera recordings, fascinated by the tinny orchestras and caricatures of voices from long ago. I was intrigued by the operatic picture sleeves, by the photographs of handsome, mustachioed men in awkward costume, by the crazed women in impractical gowns. Once, I spent hours with the encyclopedia, looking up the life story of every singer from my parents' records. It was discouraging; most of the singers were long since dead. But my personal favorite, Geraldine Farrar, the beautiful American prima donna, was still alive and living in Ridgefield, Conn., a last grand figure from opera's golden age.

So I wrote a childish love letter. "Dear Miss Farrar" it began, "although I am only 10 years old, I think your singing is wonderful...." To make sure I impressed her with my maturity, I typed the note, though this added many hours to its composition time. Barely daring to expect an answer, I mailed the finished product to her, care of RCA Victor, for whom she had recorded some fifty years before.

And she wrote back, her letter arriving on a snowy January morning—an affectionate, gentle note in her eccentric and inimitable scrawl, signed "with every blessing." Her answer went immediately into a frame, but I took it out fifteen or twenty times a day for some weeks afterwards, in order to hold the precious piece of paper in my hands.

I may well have been the youngest person in the world in love with Geraldine Farrar in 1965, but I was far from the only one. And, of course, I was nowhere near the first. For Farrar, born 100 years ago today, was America's first superstar; a world-renowned media figure in an era before movie actors had names, before the great ballplayers drew their vast audiences, before big bands and Beatlemania.

It could be argued that America had produced some important operatic figures before Farrar, most notably Maine's Lillian Nordica. But Nordica's career was pretty much behind her by the time recordings made opera into mass culture, and her few discs are, for the most part, a disappointment. Farrar, on the other hand, made many excellent records, and they brought her voice into the living rooms of millions of people who never set foot inside an opera house or concert hall.

In a time when many divas were ungainly and overweight, Farrar was slender, graceful and very pretty, and she was a legendary figure by the time she reached her mid-twenties. At the height of her fame, she had a huge, entirely unofficial fan club—the so-called "Gerryflappers"—who followed their idol from performance to performance the way rock groupies do today. Such was her prestige at the Metropolitan Opera, where she spent the better part of her career, that she was accorded her own private dressing room, a distinction which would remain uniquely hers until the arrival of Kirsten Flagstad in the mid-1930s.

Farrar grew up in Melrose, Mass.—then, as now, an attractive suburb of Boston—and became something of a local celebrity while still a child, through informal performances at churches and community centers. Her professional career began in 1901, at the Royal Opera in Berlin after studies with Lilli Lehmann. Farrar was then noted not only for her voice, beauty and what would later be called sex appeal, but for her occasional youthful audacity as well. Once, when summoned to a command performance for the German royal family, Farrar refused to wear the standard courtly black or lavender. The reason? The colors didn't suit her; she wore white. This touch of American independence apparently impressed her regal audience; the German crown prince fell in love with her, touching off a major international press scandal.

By the time Farrar returned to her native country, she was already famous. Her Metropolitan debut was in Gounod's *Romeo et Juliette* on Nov. 28, 1906. New York took Farrar immediately to heart and she quickly became the Met's stellar female attraction.

From there on her career was a series of triumphs. In 1907, she became America's first *Madama Butterfly,* in a production supervised by the composer, Giacomo Puccini, conducted by Arturo Toscanini, and featuring Louise Homer, Antonio Scotti and Enrico Caruso in the other key roles. Other highlights of her Met years were the world premiere, in 1910, of Engelbert Humperdinck's *Königskinder,* in which Farrar essayed a charming "Goose-girl" and the leading soprano role in Gustave Charpentier's *Julien,* the composer's lukewarm attempt to follow up his successful *Louise.* She was noted for her bright, vividly acted performances in the important Puccini operas, for her Manon, for her Zerlina and, especially, for her Carmen.

It was in the last role that Farrar made her moving picture debut in 1915. *Carmen* was but the first of many films the singer would make in the next few years, and it is a tribute to her acting talent that she was so convincingly able to switch her artistic medium. These were silent films, remember, and other well-established operatic luminaries, such as Caruso and Mary Garden, were stunning failures on celluloid. But Farrar built up substantial film credits, working with directors such as Cecil B. DeMille and actors such as Milton Sills and Wallace Reid. It would be illuminating to see some of Farrar's films today, but they are rarely revived.

When Farrar returned to New York, method acting now ingrained in her system, a readjustment problem came to a head on the stage of the Met. In Farrar's first post-Hollywood *Carmen*—Bizet's music restored—the soprano attempted to inject a sense of dramatic realism by roughing up the chorus girls, and by slapping Caruso, her Don Jose, full in the face. A row ensued, and the performance was nearly terminated in mid-act. Happily, things were easily patched up, to the undoubted relief of the Metropolitan management. After all, the Caruso-Farrar team was the most successful box-office pairing in operatic history, an achievement that still stands.

Farrar retired from the Metropolitan at the age of 40, with a matinee performance of Leoncavallo's all-but-forgotten *Zaza* on April 27, 1922. Although she could truthfully say that she was leaving opera because she wanted to, it was, nevertheless, time to go. She had been somewhat reckless with her voice and there were already unmistakable signs of age and wear; a private recording made only five years after her retirement reveals the technique still impeccable but the vocal bloom completely gone. The last Met performance was one of the most emotionally charged operatic events of the century; tickets sold out immediately and the scalpers asked—and received—up to 200 pre-inflation dollars per pair of tickets. After the opera was over, Farrar was mobbed by thousands of admirers who escorted her open car up Broadway and swept the retiring diva away in victory.

After her retirement, Farrar concertized for a few years, then settled quietly in Ridgefield, where she would spend the rest of her life.

Although she was widely rumored to be romantically involved with Toscanini at one point, her one marriage was to actor Lou Tellegen. Because it lasted only a short time before a bitter divorce, Farrar might have been a candidate for the life of lonely recluse. But this was not to be the case; an ardent Republican, she remained active in civic affairs, occasionally accepting speaking engagements, and for a time she co-hosted the early Metropolitan Opera broadcasts with Milton Cross. In addition, Farrar wrote poems, songs and two autobiographies. The latter of these works, *Such Sweet Compulsion* (1938) reveals a strong character, sharp insight and one of the strangest literary devices ever employed in a biography; it is purportedly only half-written by Farrar, the other half dictated from the grave by her dead mother. Despite its schizophrenia, this endearing volume is a good addition to any collection of Farrar memorabilia.

Farrar's real immortality lies in her recordings. The earliest date from 1904, made at the height of her success in Berlin. These discs, made for the Gramophone and Typewriter Company, capture the Farrar voice at its freshest, although her singing would grow more subtle as time went by. After her Met debut, Farrar recorded only for Victor and, considering the discomfort caused by the primitive recording techniques of the day, she managed to be quite prolific. She recorded several excerpts from Gounod's *Faust* with Caruso, Scotti and Marcel Journet. This group of discs, over an hour of them, made in various sessions around 1910, represents the nearest thing we have to a complete re-corded opera from the Met's golden age. Some time later, Farrar also recorded a sizable chunk of *Carmen* with Giovanni Martinelli, and a good deal of both *Madama Butterfly* and *La Bohème*. In addition, she recorded numerous songs, excerpts from *Königskinder* and some charm-ing snippets from some of the operas of Ermanno Wolf-Ferrari, all full of delights for the vocal music aficionado.

Farrar's voice was one of great warmth; her high notes had a brilliant, gemlike quality. There is a serene confidence to her singing, a subliminal awareness of her splendid technique and the seamless skein of her register. Hers was startlingly original singing, full of verve and passion, yet very modern, with none of the swooping into notes so prevalent at the time. Her singing always had a sense of center, a sense of control, which is one of the reasons her discs are more enjoyable to listen to today than those of many of her contemporaries.

Geraldine Farrar died in Ridgefield on March 11, 1967, less than two weeks after her 85th birthday. At the time of her death, the critics revived all the old clichés: an era had ended, she was the last witness from the golden age and so on. In her case, every word rang true.

The New York Times (1982)

11

A CONVERSATION WITH PHILIP GLASS AND STEVE REICH

Steve Reich and Philip Glass, two of the era's most influential composers, have had a long and tangled relationship. In 1978, just as their work was beginning to build a popular following, Columbia University's WKCR-FM produced a 20-hour festival of the music of Reich and Glass. It was the summer between my junior and senior years and, although I did not yet know it, this was the start of what would turn out to be my career. My co-producer, the late Taylor Storer, and I had a vision for WKCR and, by extension, for radio in general. We believed that there was an audience for new music on the air and, as we blasted our strange sounds into the New York night, and answered the enthusiastic, if confused, calls from well-wishers, we knew we were right.

During the festival, Reich and Glass visited the station to talk about their music. The atmosphere, charged at first, quickly grew warm and congenial. After the interview was over, we discovered that our engineer had threaded the tape recorder incorrectly and our program apparently had been lost to posterity. Almost a year later, however, a friend located a tape he had made of the program, the interview was transcribed and portions were soon published in an obscure arts magazine called Cover.

This marks what may have been the last time Reich and Glass ever sat down for a dialogue about music: The media teaming of their names, alluded to in this interview, continued until it became difficult for both men, who now take pains to dissociate themselves from one another.

TIM PAGE: How did the two of you come to know each other?

STEVE REICH: I met Philip at the Juilliard School of Music in 1958. We were both composition students and both of us had come from an undergraduate background in philosophy, strangely enough—Philip at the University of Chicago and I at Cornell University. I remained at Juilliard from 1958 through 1961, at which point I left to go to California to study with Luciano Berio. In 1965, I returned to New York City and began performing with my first ensemble; at that point it had no name. It was simply me and Art Murphy, a friend from Juilliard, and Jon Gibson, a woodwind player who now plays with Philip and whom I knew in California. So many of these paths cross!

PHILIP GLASS: We met again after I returned to New York from Paris, where I had been studying with Nadia Boulanger. We met on the occasion of Steve's 1967 concert at the Park Place Gallery, which was a very famous concert in its day. We each had been developing our own music in our own distinctive ways, and when I met Steve, it was reassuring to discover another group of musicians working in a manner similar to my own.

For a number of years immediately after that, we spent a good deal of time together. We showed our music to one another. There was a very active dialogue going on.

This is really a very small world we're talking about; a lot of these people know each other. Jon Gibson has played with Terry Riley, La Monte Young, Steve, me and with his own ensemble. Ten or twelve years ago, this phenomenon was so underground that it was really our isolation from everyone else that threw us together. It was this isolation, this feeling of banding together to escape the indifference and hostility of the outside world that created a kind of community. From that point of view, it was a very exciting and healthy period.

PAGE: Was your music influenced by your association with visual artists? I know that both of you have many artist acquaintances, and you both live in downtown Manhattan. And, Steve, every time I see a picture of a performance of your "Pendulum Music," I recognize the artist Michael Snow as one of the players.

REICH: I've been using that same photo for many years now; that's a piece that's gone out of my repertory.

PAGE: Phil, you once worked on a collaboration with Richard Serra, and Chuck Close's gigantic portrait of you fills up most of a wall at the Whitney Museum.

GLASS: Well, you must remember that we were all neighbors, and that this was as much a social environment as an artistic one. The artists were some of the first people that really supported this music—either by direct gifts to the composer, or by helping to put on concerts, or by making a loft available or by making posters. It was the first community that really put its weight behind this music. But I don't know if we can talk about influences so much as shared interests. For instance, while Steve was writing an essay entitled "Music as a Gradual Process," several well-known artists were busy creating what they called "process pieces." This wasn't a coincidence; this was a shared interest, even a shared obsession.

In those days, Lower Manhattan was a mixed environment of dancers, sculptors and musicians, all able to discuss each other's work in a fashion that was not at all academic or historical but real, alive and very rewarding. I don't think it does much good to drop names and mention all the artists who we knew in those days, but, in a sense, we all grew up together.

REICH: In the very early days, there was so little work that the number of musicians involved in playing my music, and in playing Philip's music often overlapped completely. We'd often find ourselves in the same places a night or two apart—in different lofts downtown, and then later at the Museum of Modern Art, the Kitchen or the Walker Art Center in Minneapolis. And slowly we built an audience.

GLASS: We weren't chasing after the same old gang that traditionally attends every music event in New York; we just wanted to find some people who liked our work. And we found this initial support among the artists. They didn't feel that they had to be opposed to some thing which might threaten their academic standing; they were just supporting music they liked.

REICH: Now the following has really grown. I was just out at the University of Illinois at Urbana—really out in the heartland of America. And the students came up to me and asked me genuinely intelligent and informed questions about my music, about Phil's music, about Terry Riley's music. They're talking about new music a lot, and swapping recordings, and trying out their own ideas. And I'm awfully glad, because if my music wasn't of interest to young composers, I would have failed in some fundamental way.

PAGE: Your music is so distinctly yours that it's hard to codify. It can't really be defined by a technique, the way a lot of the 12-tone composers' work can be. I wonder if you will find a group of disciples.

REICH: I don't know, Tim. I do hope that we have evolved some real, certifiable compositional techniques which are quite adaptable and should be of use to a new generation of artists. I would hope that we are deriving a new musical language.

PAGE: It's interesting that people have always seemed to lump your names together, whereas in reality your paths have greatly diverged from what were somewhat similar beginnings. How do you feel about this continued pairing of your names?

GLASS: Well, as the paths continue to diverge, it becomes easier for us to be in situations like this dialogue. It's become abundantly clear, as it always was to both Steve and me, that we are two different personalities. But a decade ago, when you had something like my "Music in Fifths" on the one hand, and Steve's "Four Organs" on the other, there wasn't a lot of other music around to compare it to except the work of Charles Wuorinen and Mario Davidovsky. In the context we sounded like Siamese twins. Ten years later, with not only ourselves but a whole new generation of people working with repetition, it is obvious that our personal ways are very distinct. I don't think we ever worried about any confusion, but other people may have.

REICH: There are obvious similarities. But anyone living in a particular period of time in the same general geographical area will probably give off a similar response if his receptors are in order. We heard the same sounds: the classics, jazz, pop and early rock. I once said that the three basic influences on my work were Bach, Stravinsky and be-bop jazz. I still feel the same way, although I've since studied African drumming and various other non-Western musics as well.

PAGE: One of the most important effects you've had on young composers has been the spiritual liberation from the old, atonal, post-Webern route which has become so institutionalized.

GLASS: Can you believe that was already institutionalized when we were students twenty years ago?

REICH: As a matter of fact, the academics hadn't even gotten to that yet. While we were at Juilliard, the blooming aesthetic was Americana. Elliott Carter's second string quartet received its premiere performance while we were there and it was considered very questionable by some members of the faculty. Some liked it, some didn't, but it was on the very edge of respectability. And if you were composing 12-tone music, you were a pariah.

PAGE: When I was studying at music school, we simply assumed that tonality was dead, repetition was frowned upon, and that steady rhythmic pulses were, after Stravinsky, played out. And then along the two of you came and turned the rules upside down. It must have taken a great deal of courage to play this music in public at a time when all of the prevailing musical trends were so diametrically in opposition.

REICH: I remember when Michael Tilson Thomas and I played "Four Organs" on an otherwise typical Boston Symphony Orchestra program at Carnegie Hall in 1973. The subscribers came to hear the other music—C.P.E. Bach, Mozart, the Bartok "Music for Strings, Percussion and Celesta" and the Liszt "Hexameron." There was a pretty full house and, at times during my piece, I would say that well over three-quarters of the people were not just booing but *really* enraged—shaking umbrellas, you know, so loudly during the piece that, on stage, we began to lose count. "Four Organs" is a piece that calls for an awful lot of concentration on the part of the performers. There was so much active feedback from the audience that we got lost, and Michael had to shout out numbers so that we could know what bar we were in. When the piece was over, a small crowd was bravoing and a much larger crowd booing just as strongly as possible. And the reactions of the press! "Primitive" was one of the kinder epithets.

PAGE: Like "Rite of Spring" all over again.

GLASS: We've all had our "Rites of Spring." It's very interesting that one can still write music that causes this kind of reaction, but I should quickly add that it's hardly been our purpose in composing. One of the things that triggers this negative reaction is a collective feeling on the part of the audience that, for some reason, they've been put on; that we're trying to make fools out of them. That's what really gets them upset. And, of course, this is the furthest thing from our minds.

Many of the problems our music had were economic. Funds for new music in America are still so limited that as soon as people begin to appear who can challenge the traditional funding of new music and raise questions about which composers should be funded, the newcomers become the young Turks and deeply threatening to an older generation.

REICH: I think you're right. There's also an element of psychology involved, because there are bound to be strong first impressions made of our music, particularly among the people who make up the faculties of music departments at the typical university. This music was alien to what was the accepted norm—even the accepted avant-garde norm—in the late 1960s which, as you suggest, Tim, was a tossup between European serialism and a certain sort of American aleatory music, both of which came out sounding atonal and nonrhythmic. This music is very tonal, very rhythmic, and it must have seemed quite frightening. It takes a long time for new music to gain acceptance—it's a pretty slow-moving art field compared to, say, painting and sculpture which, at least during the 1960s, moved very quickly. On the other hand, there does seem to be a little more permanence in the field. Once something gains the interest of other musicians and composers, there is a possibility that they will remain interested in it.

PAGE: The two of you have helped revive the tradition of composers as performers.

REICH: Well, it wasn't entirely a dead tradition. Stravinsky conducted and played a lot of his music, and made most of his income that way. Aaron Copland is a conductor. I've been told that Paul Hindemith never wrote a part for any instrument that he couldn't play himself. Bela Bartok was a concert pianist. But it is true that both of us have founded our own ensembles and that we believe in performing our own works.

PAGE: You have different attitudes towards the publication of your scores. Several of your works are published, Steve, while you, Philip, are very reluctant to part with any of your music.

GLASS: I don't want to talk for Steve, but my motivation has always been mainly economic. We both make a living as performers of our own music, and I'm unwilling to let my pieces go one at a time. I want some publisher to take the whole ball of wax; when I strike up a deal, I want to make it with one company for everything.

REICH: At this point, I have decided to publish only pieces that can be played fairly easily. Something like "Clapping Music" can be performed an awful lot more easily than a piece like "Drumming." And I'm not going to release a work which would interfere with the living I make as a performer. But some of the pieces I've published are currently out of my repertory—"Four Organs" for instance—and I would like other musicians to continue playing these works, even if I no longer do so.

GLASS: If I'm the only one who owns the music then anyone who wants to hear it has to hire me to play it. That's what it comes down to. In England, where I did a tour after several years of bickering with various sponsor groups, I was able to organize concerts for the simple reason that there was no other way for the British to hear my music. They would have been much happier performing it themselves, but they couldn't get the scores. So they had to hire the ensemble. Simple as that. I have to make a living.

REICH: I would say that the largest source of my income is European, and that's been true for a number of years. I couldn't live in New York without Europe, and yet I've never been tempted to become an expatriate.

GLASS: I'd say that 90% of my income comes from Europe. There is much more official support for the arts there than there has ever been in the United States.

REICH: Well, let's give credit where credit is due. There *is* a New York State Council on the Arts and a National Endowment for the Arts. I have been to a lot of premieres that wouldn't have been possible without their support. I certainly want to give them credit; I just think that they stand alone.

PAGE: Have you had any trouble with any of the Marxist groups over in Europe?

GLASS: None at all. I don't know about you, Steve, but, oddly enough, I've gotten a lot of support from the Left, and I've never understood it because my work is hardly political. Steve wrote some political works—like the early tape piece "Come Out," but neither of us have ever been flagwavers.

REICH: I think that there may be an assumption that we are men of the Left because, particularly in France but actually throughout Europe, our music was introduced by the so-called underground channels that were closely aligned with some of the leftist newspapers. So we became, in the eyes of a certain percent of the public, leftists. It irritates me that such political judgments can be made about artists on such flimsy evidence.

PAGE: I was just wondering if any of the more vociferous Marxists criticized you for the total control one finds in your music.

GLASS: It's happened, but it's a false issue. You could say the same thing about Webern.

REICH: Or Bach.

GLASS: Or most other composers. I have no philosophical prejudice against improvisation. It's fine; I just don't happen to be very good at it, and therefore I don't put myself in a situation where I would make an ass of myself. I admire the work of people like Cecil Taylor or Ornette Coleman or Anthony Braxton. I could go on and on, but I don't think that their use of improvisation necessarily says anything about their personal politics either.

REICH: Composition should not be used as an ideological tract. Nobody complains about the discipline in Balinese gamelan music. I've worked with Balinese musicians and if you make a mistake with them, they'll simply grab the mallet out of your hands. And they won't hit you with it, but they might like to.

GLASS: I remember trouping up from downtown with you to attend a class in African drumming. It was fascinating because the standards of performance were so far above anything I could aspire to.

REICH: The master drummer would play a pattern and we would have to play it back. And the standard of imitation was extremely high. You did it right or you did it wrong, and if you did it wrong you were corrected. Now I don't think that the political Left or champions of indeterminate music would ever dare criticize Balinese gamelans or Ghanaian drummers, but they are quite content to blast through-composed Western music. And I think that's a little odd, and well worth pondering.

PAGE: So minimalist music finds its roots in many different places.

GLASS: I want to say right now that this is a misnomer. I do not write "minimalist" music. I think that word should be stamped out. To call it minimal is just a mistake. This technique is capable of supporting music of remarkable richness and variety. If you want to call a work like "Music in Fifths," minimal, that's O.K. I guess; if you want to call "Violin Phase" minimal, that's all right too. But to apply the same term to a work like *Einstein on the Beach* or "Music for 18 Musicians"—works which are the living continuation of the ideas which first found expression in "Violin Phase" and "Music in Fifths" then the word is simply wrong.

PAGE: But nothing seems to be a suitable word! What do you call your music?

GLASS: That's a problem that's plagued all of us for years. If there had been an appropriate name, we would have leapt on it a long time ago. For this reason, we talk around the subject in various ways. I speak about music based on process, or music with repetitive structures— I think the latter name comes closer to summing this music up than anything else, because anyone who wants to talk about my work seriously is going to have to think about repetitive structures—both harmonic and rhythmic. The word "minimalist" applies only to one rather short-lived period in this music's history.

REICH: I don't think that these questions are usually answered by composers. For example, Arnold Schoenberg wanted to call his music "pantonal."

GLASS: He did?

REICH: Yes, he did. He desperately wanted to call it "pantonal." But everybody else called it atonal music and that, as they say, was that. Schoenberg didn't believe there was such a thing as atonality; you could deeply offend him by using the word.

But finally the decision wasn't in his hands. I think that the various media—whether scholarly journals, newspapers or something in between—really make these decisions. All we can do is try to convey accurate information about our work, in programs, liner notes and interviews. But we will not create the catch phrases. That's left to other people.

12 PHILIP GLASS

One two three four
One two three four five six
One two three four five six seven eight

From this unpromising beginning—a succession of numbers chanted by a small chorus beneath a stage flooded with light at the opening of *Einstein on the Beach*—has grown the most successful opera career of any composer within recent memory.

Consider. In the past dozen years, Philip Glass has created four major operas—*Einstein on the Beach* (1976), *Satyagraha* (1980), *Akhnaten* (1984) and *The Making of the Representative for Planet 8* (1988). He has also composed several chamber operas—*The Photographer* (1982), *The Juniper Tree* (with Robert Moran, 1986) and *The Fall of the House of Usher*, heard in Cambridge, Massachusetts and at Louisville's Kentucky Opera in May and June. In addition, he contributed the Cologne and Rome sections of Robert Wilson's opera *the CIVIL warS: a tree is best measured when it is down*. Most recently, he has been commissioned by the Metropolitan Opera to write *The Voyage*, a celebration of the 500th anniversary of Columbus' discovery of America.

Moreover, Glass is a genuinely popular composer. Opera companies are notoriously timid, and most modern works are studiously ignored; when a composer is lucky enough to get a production, it is usually at a small house, and after the first run, the work vanishes from the repertory. But Glass' operas rarely play to an empty seat, and there have been several separate productions of *Satyagraha* and *Akhnaten*, with a new version of *Einstein on the Beach*, directed by Achim Freyer, promised for the Stuttgart Opera this October.

Glass and his performing ensemble now present some ninety concerts a year and are capable of selling out Carnegie Hall one night and a Midwestern rock club the next. Conservatory students diligently analyze the composer's unusual orchestration, while their more hedonistic contemporaries are content to blare Glass albums from dormitory stereo systems.

Not surprisingly, Glass' commercial success has not sat well with some of his more conservative colleagues. ("Glass is not a composer," one of them told me. "That's all. He's simply not a composer.") Nor has his music been hailed universally by the critics. "[Glass' operas] stand to music as the sentence 'See Spot Run' stands to literature," Donal Henahan wrote in the *New York Times* after the first New York City Opera performance of *Akhnaten.*

And there have been some important critical reversals along the way. Andrew Porter of *The New Yorker* wrote rapturously of *Einstein on the Beach* and *Satyagraha* when they were new, dismissively thereafter. ("My own responsiveness to minimalism in opera—to minimalism of all kinds—soon diminished," he explains in a connective passage after a reprint of his paean to *Satyagraha* in his latest book.)

Others have felt very differently indeed. "One listens to the music, and somehow, without quite knowing it, one crosses the line from being puzzled or irritated to being absolutely bewitched," Robert Palmer of the *New York Times* has written. "The experience is inexplicable but utterly satisfying, and one could not ask for anything more than that."

I count myself an admirer. More so, perhaps, than any other composer of our time, Glass has fashioned his own inimitable aesthetic. Those who disdain Glass' work for its seeming simplicity miss the point. Of course it is simple. But it is not easy, and it is very difficult to imitate. What may impress one as a banal chord progression at the beginning of the piece grows increasingly interesting as the work progresses, and as we examine it from each new vantage point that the composer presents to us. This sort of musical alchemy is what sets Glass apart from his many followers.

Glass himself doesn't care what the critics say. "Don't tell me whether the review was good or bad, tell me how much space the paper gave to the event," he said one afternoon while relaxing in the basement of his elegantly funky town house in Manhattan's East Village. He had been up since five in the morning and working since dawn to fulfill his quota of music for the day. Yet he was full of energy and seemed ready to reminisce, gossip and philosophize.

"Only a few people read reviews through, you know, and only a few of *those* people really care what the reviewer thought," he continued. "I had a vivid demonstration of this. A cousin of mine in California sent me a long review of my work in a local paper. And the letter said, 'I was so pleased to see this review, and I'm so proud of you, Cousin Philip.' And it was one of the worst reviews in history, just nothing good about it at all. So I wondered what she could have been thinking of, sending it to me, and then I realized she hadn't read it and was just happy to see me on the page. That was all that mattered."

Glass is amiable, articulate, unpretentious and funny, with a hint of defiance toward the musical establishment. "We're getting ready for the fourth production of *Akhnaten,* which is going to take place in Brazil," he said. "You just wait, this is the opera that everybody is going to want to do, despite the initial response. I'd really like to have it performed in Egypt, like *Aida,* among the pyramids"—*Akhnaten* is the study of an eighteenth-dynasty pharaoh—"and we were close to making some progress. But no luck. They'll wait 100 years, and then they'll get around to it."

Though he loathes the term, Glass is often classified as a "minimalist," along with such fellow composers as Steve Reich, Terry Riley and John Adams. His mature music is based on the extended repetition of brief, elegant melodic fragments that weave in and out of an aural tapestry. Listening to these works has been compared to watching a modern painting that initially appears static but seems to metamorphose slowly as one concentrates. Compositional material is usually limited to a few elements, which are then subjected to a variety of transformational processes. One shouldn't expect Western musical events—sforzandos, sudden diminuendos. Instead, the listener is enveloped in a sonic weather that twists, turns, surrounds, develops. Detractors call it "stuck-record music" and worse, but Glass has brought an excited new audience into the opera house.

Born in Baltimore in 1937, Glass began his musical studies at the age of eight. At fifteen he entered the University of Chicago, where he majored in philosophy but continued what had already become an obsessive study of music. After graduation, he went the route of many other young music students—four years at the Juilliard School in New York, later work in Paris with the legendary pedagogue Nadia Boulanger, who had taught Aaron Copland, Virgil Thomson, Roy Harris and other American composers. During his time in Europe, Glass also was exploring less conventional musical venues, working with Ravi Shankar and Allah Rakha. He acknowledges non-Western music as an important influence on his style.

In 1967 Glass returned to New York City, establishing himself in the blossoming downtown arts community. "At first, my compositions met with great resistance," he said. "Foundation support was out of the question, and the established composers thought I was crazy. I had gone from writing in a gentle, neoclassical style that owed a lot to Milhaud into a whole new genre, and the timing wasn't right."

So Glass worked as a plumber, drove a cab at night and spent his spare time assembling an early version of the Philip Glass Ensemble. The group, which consists of seven musicians playing keyboards and a variety of woodwinds, began concertizing regularly in the early 70s, playing for nothing or asking for a small donation. "People would climb six flights of stairs for a concert," Glass remembered. "We'd be lucky if we attracted an audience of twenty-five, luckier still if half of them stayed for the entire concert." Then as now, audience response was mixed. Some listeners were transfixed by the whirl of hypnotic musical patterns the ensemble created, while others were bored silly, hearing only what they considered to be mindless reiteration.

But slowly, very slowly, the concerts gained a cult following, and then suddenly *Einstein on the Beach,* a collaboration with the austere theatrical visionary Robert Wilson, made Glass famous. *Einstein* broke all the rules of opera. It was five hours long, with no intermission—the audience was invited to wander in and out at liberty during performances. Glass' text consisted of numbers, solfège syllables and nonsensical phrases by Christopher Knowles. The Glass-Wilson creation was a poetic look at Albert Einstein: scientist, humanist, amateur musician— whose theories led to the splitting of the atom. The final scene depicted nuclear holocaust: With its Renaissance-pure vocal lines, the blast of amplified instruments, a steady eighth-note pulse and the hysterical chorus chanting numerals as quickly and frantically as possible, this was a perfect musical reflection of the anxious late 70s.

Einstein was presented throughout Europe, then at the Metropolitan Opera House for two performances in November 1976. But Glass lost a great deal of money on the production. "In the winter of 1976–77, what we had come to refer to as the 'Einstein debt' seemed a huge weight that could never be rolled away," Glass recalled in his book, *Music by Philip Glass.* So he returned to driving a cab while working on his next opera, *Satyagraha,* produced in Europe in 1980 and at the Brooklyn Academy of Music in 1981.

Satyagraha, a metaphorical portrait of Gandhi, was completely different—closer to religious ritual than entertainment, to mystery play than to traditional opera. While *Einstein* challenged all received ideas about what opera, even avant-garde opera, should be, *Satyagraha* fit Glass into the mainstream. Where *Einstein* had broken the rules with modernist zeal, *Satyagraha* adapted the rules to the composer's aesthetic—a much more difficult task.

Akhnaten was another step in the same direction. Here there were genuine set pieces—duets, ensembles, choruses and one long, challenging aria for countertenor—and the work was scored for a more or less conventional orchestra. Though it sold out all performances at the Houston Grand Opera and New York City Opera, it was widely considered a failure, in part because of an ugly, pretentious production that resembled a "*Saturday Night Live*" sketch, along the lines of "The Coneheads Go to Egypt." (A recording has been issued since, to better press.)

Still, it was an unusual work—the language was ancient Egyptian, and the only English in the score was narration. As with *Einstein* and *Satyagraha*, there was little drama in the sense that Verdi or Puccini might have understood it. Which suits Glass fine. "I don't really take opera composers as models," he said. "Isn't that the point? I like to listen to the same works that everyone likes to listen to—Mozart, Rossini, Verdi, Wagner and the rest. But it never occurred to me to write like any of them, any more than it would to put on the clothes of a nineteenth-century Italian. These works occupy a world of their own, and it is a beautiful world, one that enriches us all. I am a devotee of museums, but I don't want to live in one."

Glass names Virgil Thomson as one of the predecessors he most admires. "He's a friend of mine. We were talking not long ago, and he said, 'People like us'—I was flattered that he included me—'People like us, we're theater composers. We might write an occasional piece of abstract music, but we're really theater composers.' And he's right. I consider my three—now four—big operas my most important works.

"I'm not so interested in relating a story in the traditional sense of the word. Even *The Fall of the House of Usher*—you know, you can tell the plot in a few sentences. A brother and a sister are living alone in the ancestral home. A friend visits. The sister, quite ill, dies, or so it seems. She is buried in the vault, prematurely as it turns out. She comes back and murders the brother. The visitor runs from the house as it collapses. That's it. But for me it is a story that gives scope for an emotional examination of Poe's world. My score is eighty-five minutes of musical atmosphere with a simple tale at the bottom of it.

"In my theater pieces I like to leave some passages in which the eye can wander. In *Usher* there is a six-minute scene with only three lines. Not much happens. One person hears a voice, and a butler brings in chairs. But the music and the staging should keep the spectator's attention."

Glass is particularly excited about *The Making of the Representative for Planet 8*. "It's my first big opera in English," he said. "Like the others, it's an opera of ideas. The plot is simple. It's the story of a planet that is entering an ice age through what Lessing calls a 'cosmological disaster.' It's basically about a race of people who are about to die, about a planet that is losing the heat of the sun. It's one of the saddest stories you can imagine, and a lot of people have treated the book as if it were science fiction, but it's really an allegory.

"The power of allegory is that we can talk about things that are difficult to talk about, things that are close to us objectively. You know, it really isn't happening here, but on Planet 8. If we actually set the story on Planet Earth, it would be too depressing to think about. Nothing much really happens in the opera. People face their death, the death of their species. But there are many different ways of dying. An allegory, really, almost like a Biblical story.

"The emphasis of my work has been on collaboration throughout, whether with Bob Wilson or Doris Lessing, whether in music theater, film or dance. I'm convinced that this is one of the major reasons that I followed a path different from other composers. There was always input from another person."

Glass expounds on this in his book:

> For the most part, [the standard] operas in the Italian and German traditions were the work of one man with one vision (the contribution of librettists notwithstanding). The opera houses of the past simply produced these works and did not function as workplaces where artists from different fields collaborated on joint projects. Most modern operas written for present-day opera houses are conceived in exactly the same way. It is not so odd, then, that music theater works coming from a new and very different tradition should be greeted with surprise and even alarm— when they are acknowledged at all—by most producers of traditional work.
>
> Still, I feel very positive about our inherited and changing world of music theater. New works not modeled on the past are being created, producers are beginning to appear in cities and countries all over the world, and the public for these new works is very much there. When new works can outsell classics of the Italian and German repertory, as is actually happening all around us, it is hard not to sense a growing momentum toward a new and revitalized future for the music theater of our time, and I think this will have the proper effect on our opera houses. I don't doubt that the world of traditional repertory opera will eventually be dragged—probably screaming—into the twenty-first century, and that will be a whole new story.

Glass' current projects include the score to *Powaqqatsi,* a sequel to the acclaimed film *Koyaanisqatsi* by Godfrey Reggio; a ballet for Molissa Fenley; a science-fiction music drama with David Henry Hwang, *1,000*

Airplanes on the Roof; and a new opera with Robert Wilson, *The Palace of the Arabian Nights,* which will receive its world premiere at the Théâtre du Châtelet in Paris next December.

Beyond that lies *The Voyage,* Glass' major project for the Met. When the commission was announced in March, its fee of $325,000 was called the highest ever paid for an opera. Glass notes that "A lot hasn't been decided yet," but he does cite the theme he has chosen, "the idea of the great explorers at different times in our history. There's fantasy and history in it. By the time I get done with it, it will be a pretty good tale." He is scheduled to deliver the score in 1991 for production the following year. The Met management, which has been criticized both for essaying so few new works and for offering any at all, discounted the current vogue for Glass as a determining factor. General manager Bruce Crawford said, "The synopsis caught both Jim [James Levine] and me and produced our enthusiasm," though the composer's popularity "didn't hurt" and showed his capability to deliver a viable score.

Glass is a resolutely cheerful man. He now lives with his girlfriend, the artist Candy Jernigan, in a guacamole-colored house on a particularly seedy corner of the East Village. Glass has two children from a former marriage: Daughter Juliet is off at college in Oregon, son Zachary is living at home, studying music. "He's not interested in my stuff at all right now," Glass said with a grin. "It's probably healthy."

Jernigan's art, on prominent display, is good-humored and allusive, occasionally gently punning (a drawing on the wall features the labels of several varieties of canned Spanish beans, straight from the shelves of a local *bodega;* it is called "Homage to Goya.") There are books by the photographer Robert Mapplethorpe on the table, also a volume of illustrated Talking Heads lyrics to which Jernigan contributed. There's even something called *The Wonderful Private World of Liberace,* a coffee-table pictorial celebration rather than an exposé. The phones ring constantly. A cat lies on a counter. The atmosphere is friendly and informal. Glass bought his house in 1984 and, when he is in New York, he leads a quiet, strictly disciplined existence. "I work best in the early morning," he said, "so I get up around five or six and compose until noon or so.... Occasionally I will have to write really quickly. I once wrote and recorded forty-five minutes' worth of incidental music in a week—*literally* a week. The guys in the studio were working three hours behind me. I'd finish a movement, and they'd learn it and record it. I do compose quickly, you know, but nobody likes to work *that* fast. Still,

I liked the incidental music enough to fashion a twenty-minute orchestral suite from it. It's not bad at all.

"Just recently, somebody asked me how long it took me to write one of my pieces he'd just heard. And I said about forty years. I know it sounds like a cynical answer, but I've been working at this business of composing since I was eight. And everything adds to the whole—everything I've learned influences the music I write.

"I deal with business matters in the afternoon—recording, auditioning, the *real world*. Giving interviews, too. I decided long ago that I was my own best spokesman, and I made it a point then and there to talk with anybody. I'm rather proud of the fact that I've spent as much time talking with reporters from high school newspapers as I have talking with *Time* magazine. Kids grow up, you know, and these same folks who are now at high school newspapers or college radio stations have a habit of becoming music professionals. It's not just that I like to talk with young people—which I do, and sometimes they are better prepared than professional critics—but it's good business as well." Evenings are reserved for friends, for family, an occasional party, even a concert or two.

Of course, Glass isn't home very often these days. He spends much of the summer in Nova Scotia, where he devotes himself to a strict regimen of composing. And then there are the tours. "I cross the country twice a year with the ensemble," he said. "I play all over the place. I've played in six cities in Montana. Did you *know* there were six cities in Montana? Do you know the population of Laramie, Wyoming? I do. I know where the pizza places are. I know both of the hotels."

Glass leaned back and grinned. "You know, it's just like I always thought it would be. When you start to be well-known, you get to play all the really *small* places."

(1989)   (Portions of this article appeared in Opera News *and* Newsday*)*

13 GLENN GOULD: A REMINISCENCE

We were speeding through the empty streets of Toronto in Glenn Gould's black Lincoln Continental. It was three o'clock in the morning and "Toronto the Good"—now grown to Canada's largest city but still called by its Puritan nickname—was long since fast asleep. But Gould was awake, skillfully maneuvering the heavy automobile down the deserted highway. Although it was a mild night in late August, he was dressed in two sweaters, woolen shirt, scarf, gloves, coat and slouch hat, cheerfully conducting a tour of his lifelong home, his chosen Elba.

This was the second day of my visit and I was trying hard to reach the airport on the car phone. For when Gould had finally agreed to give me a personal interview, I had deliberately limited the length of my trip, not wanting to disturb this supposed hermit with an inordinate stay. But I was enjoying myself immensely and so, to my satisfaction, was Gould. He had made me feel such an honored guest that I was trying to book a later flight back to New York, in order to continue our marathon discussion, which had begun as soon as I arrived at the Inn-on-the-Park, the suburban hotel that housed his studio.

The night before, we had talked until 4 A.M. Later on this morning, as I would reluctantly make my way off to my room, the sun probably shining outdoors—no light seeped into Gould's shuttered sanctuary— my host would still be washing down cups of weak tea, sifting through his collection of videotapes and recordings and eager for more talk of music and philosophy.

The reason for our late-night sojourn through Toronto was to tape a special promotional interview to accompany the release of Gould's second recording of Bach's "Goldberg" Variations. He had grown dissatisfied with the acoustics in his own studio, so we had spontaneously decided to continue our session in another part of the city. Our engineer, Kevin Doyle, knew of a pop music studio that would be empty, so we transferred the entire production downtown. Here Gould relaxed at the keyboard of a Yamaha baby grand piano, casually playing through his own transcriptions of Richard Strauss operas.

"Given my druthers," Gould once said, "I'd rather not live in a city at all." And indeed he could not be said to live in Toronto in the usual sense of the word. At the time of my visit, he had not set foot in a store or restaurant in some years. Aides did the shopping; other items could be delivered. His only meal of the day was taken in his studio at 6 or 7 in the morning, invariably alone and just before his bedtime. He never attended movies or concerts, slept by day and worked by night, keeping in touch with friends almost exclusively via the telephone.

It was not easy for an outsider to reach Gould. He did not live at the apartment where he received his mail, and his several phone numbers were all unlisted. In fact, even for those who had his number, he was difficult to reach. An operator picked up all of his calls (she was still answering his phone months after his death) and would patiently explain that no, Mr. Gould was not in at the moment but would certainly return the call. Which, in fact, Mr. Gould just might do.

Until my arrival in Toronto the preceding day—August 20, 1982— Glenn Gould had been one of the best friends I never met. We first spoke in 1980 on the occasion of the 25th anniversary of his contract with Columbia Masterworks. Gould, who cared for neither anniversaries nor interviews, agreed, after some persuasion from CBS, to talk to a few select journalists. I answered this open call, aware that I needed to pass a screening process. It was understood from the beginning that there would be no personal contact, that the interview would be con- ducted over the phone and that all topics would be cleared in advance.

So I purposely steered clear of the standard litany of queries— When will you make your historic return to the stage? Why do you sing when you play?—all questions which he had answered many times. I concentrated instead on his concept of "contrapuntal radio," sound documentaries in which three or more voices were mixed at a time, a musical exaltation of speech. I would ask him about composers he loved—Richard Strauss, Arnold Schoenberg, Jean Sibelius and, of course, Johann Sebastian Bach.

I didn't think I stood a chance of getting the interview—the *Soho News* had a tiny circulation, but Gould assented and we spoke for two hours one Saturday night. Unfortunately, the *Soho News* was then under the direction of a particularly acerbic arts editor; she agreed to make Gould a cover story but invented a headline that referred to him as a "pianist and crank." My article (slightly amended, it is in this volume) had set out to prove that Gould was anything but a "crank" and I left a profuse apology with his operator, convinced that my contacts with him were at an end. But he called back a few minutes later, told me that he understood the perils of journalism, and that he loved the piece. After that we were friends and he would regularly call, usually late at night, always person-to-person, and we would talk for hours.

Conversation was one of Gould's great joys. He was a superb mimic, and an inspired raconteur, one of those rare people who could spin out a story for an hour, in elaborate detail, and keep the listener's rapt attention. He was witty, kindly, energetic and intensely interested. He extended instant camaraderie to anyone whose company, telephonic or otherwise, he enjoyed. One evening, he called to sing the Brahms G minor rhapsody in its entirety, in order to make a point about tempo relations between the various sections of the work. Another night, he read through several pages of Timothy Findlay's novel *The Wars;* he had provided the music for a grave and beautiful motion picture based on the book. And once, impressed by an article I had written for *High Fidelity* about minimalist music—which fascinated him, although he detested it—he sang an impromptu "minimalist" song to my answering machine.

I had always assumed that I would never meet Gould in person and had grown comfortable with the idea. He felt that personal en-counters were, for the most part, disturbing and unnecessary, and claimed that he could better understand the essence of a person's thought and personality over the phone. He had friends throughout the world whom he had never met; his monthly phone bill regularly ran to four figures. One evening, however, I asked if I might visit him in Toronto. To my surprise, he liked the idea and even suggested a project we could do together—a radio interview about his recent re-recording of what had come to be considered his signature work, Bach's "Gold-berg" Variations.

Our supposedly spontaneous discussion was in fact carefully scripted. Gould wanted to leave nothing to chance, so he distilled a dialogue from a phone conversation we had had about the two records, all the time making sure that my opinions were correctly represented. He then sent me a copy on which to make any changes I wished. I touched up the manuscript a little but he had been remarkably faithful to my words and I felt comfortable proceeding with our radio play.

And so, two weeks later, I found myself at the Inn-on-the-Park. Within the confines of this sprawling and labyrinthine building, around the corner from the gift shop and down the hall from the discothèque, amidst the conventioneers and bikinis, Gould spent most of his waking hours for the last five or six years of his life. His studio was cell-sized— the windows blocked, the curtains drawn. There was no way of telling what time it was, what the weather was like, what the headlines might

be. The studio was cluttered—full of tapes, tape machines, video recorders, an inexpensive stereo and a television. The bathroom was lit-
tered with empty bottles of Valium—Gould, who was a strict teetotaller
and disliked cigaretttes, cheerfully admitted his dependence on tran-
quilizers, a habit which apparently dated back to his years on the concert
stage.

At the Inn-on-the-Park, he lived the life of a McLuhan-age monk:
editing his tapes, talking with friends on the phone, writing criticism
and theoretical articles, and reading philosophy and theology through
the long nights. Here he had found the seclusion he had so long desired.

Outside the city, away from the people who knew him when, the
ancient history, the memories, Gould lived a different life. Not even his
record producer knew for sure where he stayed when he came to New
York. In the winter, he haunted out-of-season spas—in depressed min-
ing towns on the northern shore of Lake Superior, or on some favorite
islands off the coast of the Carolinas. Wherever he went, the phone
went too, and he was immensely proud of the fact that, due to the
Lincoln Continental's cellular phone, he had been able to summon
emergency help for accident victims one snowy night on the New York
State Thruway.

He had created a studio at the Inn-on-the-Park in the mid-70s
after discovering that it was the only hotel in the area that offered round
the clock room service. Because he was in the habit of ordering several
pots of tea over the course of an evening, Gould greatly appreciated this
amenity. He was served by bemused waiters, who held him in obvious
affection but couldn't help but be curious about the strange, unshaven
gentleman who greeted them so formally and habitually ordered an
expensive full course dinner at sunrise. The day staff did not know
Glenn Gould at all—were indeed unaware of his presence in a remote
corner of the hotel. He was not listed on the register and one of the
first things he did after moving in was to disconnect the switchboard
phone and install one of his own.

When I had made my reservations at the hotel, he had insisted that
I stay no higher than the third floor. (There had apparently been a fire
at the Inn-on-the-Park some years before, with a few casualties, and
Gould, his hospitality tinged with phobia, was determined that his guests
were not going to be trapped in a "towering inferno.")

I was momentarily shocked when Gould came to fetch me. He
looked older than his forty-nine years, ethereal, with the air of a tired
visitor preparing to cast off his wasted body and metamorphosize as
pure spirit. Gone was the slim, mercurial, oddly beautiful young man
whose keyboard acrobatics had so dazzled audiences in the late 50s and

early 60s. In his place was a heavy-set, balding and perpetually rumpled man. It put one in mind of Dorian Gray: Nobody outside his inner circle had seen him for so many years that in the minds of many, he had remained the young man captured in the quarter-century old Columbia Masterworks publicity glossies.

And so, in a way, he had. For when he shook my hand—in marked contrast to the germophobic legend—I was immediately aware of a childlike humor and curiosity, of a mischievous streak (although he was a man entirely without malice) and of a delight in his idiosyncratic existence.

"These are the happiest days of my life," he said one evening during my stay. Nothing in particular had occurred to motivate the comment but one could understand the contentment. He seemed to have come to terms with his genius and with the gifts which went hand in hand with his gift. As his first biographer, Geoffrey Payzant, noted: "Glenn is an exceedingly superior person, friendly and considerate. He is not really an eccentric nor is he egocentric. Glenn Gould is a person who has found out how he wants to live his life and is doing precisely that."

Gould was born only a few miles from the Inn-on-the-Park, on September 25, 1932. The son of a middle-class furrier and a piano teacher, he was composing by the age of five and entered the Royal Conservatory of Music five years later. By fourteen, he had graduated with an associate degree and on May 8, 1946, Glenn Herbert Gould made his first public appearance as a piano soloist at Massey Hall in Toronto. Later, despite success around the world, he would continue to live quietly in the city of his birth. In fact, the only city he liked, despite strong reservations about its political and economic system, was Leningrad. "If I had to live in New York or Rome," he would often say, "I'd have a complete nervous collapse."

Gould's career was established almost overnight with the release of his first Columbia album in 1955—a joyous, fleet, highly original rendition of the "Goldberg" Variations. The disc quickly became a best-seller and has remained in the catalogue for more than three decades.

In 1964, after nine years of international acclaim, he suddenly announced that he was giving up live performances to concentrate on recordings. No famous pianist had ever done anything like it before. This was heresy: Gould—who had been awarded the highest possible honors for a concert artist, rave reviews and sold-out engagements worldwide—was simply walking away from it all.

Actually, the young pianist, then thirty-one years old, had reasonable explanations handy for his decision to quit the stage. He was tired

of what he called the "non-take-two-ness" of the concert experience—
the inability of the performer to correct finger slips and other minor
mistakes. "All other creative artists are able to tinker, to perfect, but
the live performer must recreate his work from scratch every time he
steps onto a stage." In addition, Gould believed that a "tremendous
conservatism" takes over any artist forced to perform the same music
again and again, until it becomes difficult, if not impossible, to move
on. "Concert pianists are really afraid to try out the Beethoven Fourth
concerto if the Third happens to be their specialty. That's the piece
they had such success with on Long Island, by George, and it will surely
bring them success in Connecticut!"

In 1964, Gould abandoned concertizing to become a recording
artist. From the beginning, Gould had hated live performances. With
his sudden fame, he had also discovered that he hated touring, hated
flying, hated the extramusical hysteria which accompanied him everywhere
(a woman, who claimed to be his wife, wrote to him almost every day
over a twenty-year span). Finally he decided that the whole business of
being a concert artist had gotten in the way of making music. "At
concerts, I feel demeaned," he complained, "like a Vaudevillian."

Indeed, the public had come to expect a circus act of sorts from
Glenn Gould. His idiosyncratic way with the piano made him excellent
copy for journalists. He had a penchant for singing along—loudly—
while he played. He himself apologized for this quirk: "I don't really
know how anyone puts up with it but I play less well without it." He
favored a very low seating—bringing along his own traveling folding
chair which set him at about eye level with the keyboard. He wore coats
onstage even during the summer—"I have an absolute horror of catching
cold," he explained—and would sometimes perform in fingerless gloves.
Those who misunderstood what one writer called his "genuine and
profound strangeness" often dismissed him as a publicity hound determined only for notoriety. There was a side to Gould that enjoyed being
outrageous but there was usually an underlying seriousness to even the
oddest things he said.

Any writer in search of concrete "events" after Gould's seemingly
abrupt but in fact long-considered decision to quit the stage will find
little to hold on to. There is the legacy of some eighty recordings:
brilliant, iconoclastic, occasionally disastrous but never less than interesting. He was a composer as well as a pianist (his youthful string quartet
was recorded by the Symphonia Quartet on a now scarce Columbia disc
and he arranged the soundtrack music for several films, the best known
of which was *Slaughterhouse Five*.) He also wrote a good deal of witty,
learned and original music criticism.

In the late 1970s, Gould began a fruitful collaboration with film-maker Bruno Monsaingeon, which would lead to three telefilms—"Glenn Gould Plays Bach"—for a German production company. In 1982, having come a mysterious full circle, he re-recorded Bach's "Goldberg" Variations, which was issued by CBS only two weeks before his death on October 4, 1982.

Gould was happy to have re-recorded the Variations. He had come to thoroughly detest the earlier version, speaking of it, in stentorian tones, as the "most overrated keyboard disc of all time." Special contempt was reserved for his early rendition of the haunting 25th variation which, he felt, was played "like a Chopin nocturne." "It wears its heart on its sleeve," Gould said. "It seems to say—Please Take Note! This is Tragedy. You know, it just doesn't have the dignity to bear its suffering with a hint of quiet resignation."

The new recording was much more sober and introspective, with generally slower tempos. Gould analyzed the tempo of everything he recorded and would often replay a work several times in the studio before choosing a final mix to release. He was annoyed by the criticism his efforts sometimes generated: "Many people have alleged that my tempi are arrived at capriciously. While one could perhaps describe them as arbitrary or even willful, to claim that they are capricious is more or less equivalent—and I use the analogy with all appropriate modesty—to declaring that Gustav Mahler's orchestration is careless."

"I do not think I will be making piano records in a few years," he said at the close of a long editing session. "When I have recorded all I want to play, I will move on to other things." Indeed, he had already begun to expand his activities—with "contrapuntal radio" and other, more traditional radio programs.

Again with no audience present, Gould had become a conductor, working with a pickup chamber orchestra he had assembled in Toronto. The first project was Wagner's "Siegfried Idyll." He had long loved the piece; in fact, he had transcribed it for solo piano and recorded it in the early 70s. The new recording, which was in final mixes at the time of his death, must be the slowest "Siegfried Idyll" in history and is of melting and surpassing tenderness: Once heard, it renders most traditional interpretations cursory. Richard Strauss' late masterpiece "Metamorphosen" was scheduled to be Gould's next conducting project but he did not live to begin it.

"The purpose of art is not the release of a momentary ejection of adrenaline but rather the gradual, lifelong construction of a state of wonder and serenity." Gould wrote these words in 1962 and they served as his credo to the end. When I think of Glenn, the first image that comes to mind is one of him at the piano that August night, as he played through the Strauss operas he loved. The studio was deserted, Toronto was fast asleep but Glenn Gould was making music and the Yamaha suddenly became a six-foot-square orchestra. Dense contrapuntal lines, translucently clear and perfectly contoured, echoed through the room. Far from the eyes and ears of the curious world, the hungry fans and disapproving critics, the lucrative contracts and percentage deals, he played through the night, lost in the joy of creating something beautiful.

(1983/1988)   *(Portions of this article originally appeared in* Vanity Fair.*)*

14 GLENN GOULD: THE PIANO QUARTERLY INTERVIEW

TIM PAGE: Glenn, it's now about seventeen years since you left the concert stage. I'm not going to ask you why you left or whether you will return, both questions that you have answered eloquently on a number of occasions. But when you quit the stage, you stated rather unequivocally that the live concert was dead, period, and that recordings were the future of music. Since 1964, however, we have seen a tremendous resurgence of interest in the concert hall—the success of such endeavors as New York's Mostly Mozart Festival is a good example—while the recording industry is in serious trouble. Any second thoughts on this subject?

GLENN GOULD: Well, I did give myself the hedge of saying that concerts would die out by the year 2000, didn't I? We still have nineteen years to go, and by that time I will be too old to be bothered giving interviews [*laughs*], and I won't have to be responsible for my bad prognosis! As to the recording industry being in trouble, I remain optimistic. I suspect this is a cyclical thing; recording is not really in trouble in those countries where classical music means a great deal—in Germany, for example. This trouble is, to a large extent, North American; it's been coming on quite gradually for a number of years now, and it may or may not reverse itself. If it does not, it simply means that Americans are not terribly interested in classical music.

On the other hand, it doesn't seem as though the concert is going away as fast as I rather hoped it would ...for the good of all mankind. It has, however, changed. I haven't been to a concert since 1967, when, under considerable pressure, I attended a friend's recital. But I get the impression that a great many contemporary concerts are like reincarnated versions of the kinds of shows that Hans von Bülow did in Toronto a hundred years ago, when he played Beethoven's "Appassionata" Sonata immediately following a trained-horse act!

T.P.: A sort of contemporary vaudeville?

G.G.: Exactly! There is a return to that "trained-horse act" type of concert where a bit of this is followed with a bit of that and then a bit of something else—which I think is actually very nice. Twenty years ago there were very few flexible chamber concerts; you had a string quartet playing Beethoven or whatever, but there was no intermingling of interchangeable modules as now exists. That's all changed; I don't know if this is a sign of desperation—that the solo act can't sustain an entire evening anymore—or simply a more imaginative way of thinking, or possibly even a complete return to the musical thought of the 1880s. I'm not sure what significance this all has.

T.P.: I know you have a dim view of concerts in general. You once told the *New York Times* that you found all the live arts "immoral" because "one should not voyeuristically watch one's fellow human beings in testing situations that do not pragmatically need to be tested."

G.G.: Yes, I confess that I have always had grave misgivings about the motives of people who go to concerts, live theater, whatever. I don't want to be unfair about this; in the past, I have sometimes made rather sweeping generalizations to the effect that anybody who attends a concert is a voyeur at the very best, and maybe a sadist to boot! I'm sure that this is not altogether true; there may even be people who prefer the acoustics in Avery Fisher Hall to those in their living room. So I don't want to be uncharitable. But I do think that the whole business about asking people to test themselves in situations which have no need of their particular exertions is wrong—as well as pointless and cruel.

I'm afraid that the "Let's climb Everest just because it is there" syndrome cuts very little ice with me ...there's a pun in there someplace. It makes no sense to do things that are difficult just to prove they can be done. Why climb mountains, or ski back down, or dive out of airplanes or race motor cars, unless there is a manifest need for such behavior?

The concert had been *replaced*, you know. I don't want to bore you with all the reasons why I think technology has superseded the concert—I've enumerated them on many other occasions, and I don't want to do that act again. But there is one reason which I think bears on this question: Technology has the capability to create a climate of anonymity and to allow the artist the time and the freedom to prepare his conception of a work to the best of his ability, to perfect a statement without having to worry about trivia like nerves and finger slips. It has the capability of replacing those awful and degrading and humanly damaging uncertainties which the concert brings with it; it takes the specific personal performance information out of the musical experience. Whether the performer is going to climb the musical Everest on this particular occasion no longer matters. And it's for that reason that the word "immoral" comes into the picture. It's a difficult area—one where aesthetics touch upon theology, really—but I think that to have technology's capability and not to take advantage of it and create a contemplative climate if you can—*that* is immoral!

T.P.: When I said the recording industry was in trouble, I was perhaps thinking too much of economics, for in a strictly artistic sense it is certainly alive and well. In recent days there have been recordings of much formerly obscure material—early Haydn symphonies, Schubert operas, lesser-known Bach cantatas—which went unheard for many years. And a lot of new works have been recorded. Let's talk about your repertoire. While you have recorded a fair amount of the standard literature—Bach, Beethoven, Mozart, etc.—you have avoided recording some of the standard piano composers. For instance, do you think you will ever make a Chopin record?

G.G.: No. I don't think he is a very good composer. I played Op. 58 when I was younger, just to see how it would feel. It didn't feel very good, so I've never bothered to play any more Chopin.

I have always felt that the whole center core of

the piano recital repertoire is a *colossal* waste of time. The whole first half of the nineteenth century—excluding Beethoven to some degree—is pretty much of a washout as far as solo instrumental music is concerned. This generalization includes Chopin, Liszt, Schumann—I'm tempted not to say Mendelssohn, because I have a tremendous affection for his choral and chamber works, but most of his piano writing is pretty bad. You see, I don't think any of the early romantic composers knew how to write for the piano. Oh, they knew how to use the pedal, and how to make dramatic effects, splashing notes in every direction, but there's very little real *composing* going on. The music of that era is full of empty theatrical gestures, full of exhibitionism, and it has a worldly, hedonistic quality that simply turns me off.

Another problem as I see it is that Chopin, Schumann and company labored under the delusion that the piano is a homophonic instrument. I don't think that's true; I think the piano is a contrapuntal instrument and only becomes interesting when it is treated in a manner in which the vertical and horizontal dimensions are mated. This does not happen in most of the material written for it in the first half of the nineteenth century.

In the late romantic period lies the big tragedy, for the composers in that period—Wagner, Richard Strauss, possibly Mahler—those composers who could have written with a tremendous penetration of the intermingling of harmonic and thematic language just basically chose not to write for the piano at all. Wagner wrote an early sonata, but it makes Weber look like one of the great masters of all time by comparison. I suspect that Wagner had no real understanding of the piano, for the accompaniments to the *Wesendonk Lieder*, which are fine in their orchestral arrangement, don't work well on the piano at all. I transcribed and recorded a few of Wagner's large orchestral pieces some years back. It was a real labor of love; I simply wanted to have something of Wagner's I could play.

On the other hand, I have been recording the early Richard Strauss piano works—Op. 3, Op. 5, pieces

Strauss wrote when he was sixteen—and they are minor miracles: as refined, as polished, as anything Mendelssohn did in his teenage years. And with the exception of Mendelssohn, no sixteen-year-old has *ever* written with such craft and assurance—I am *not* forgetting Mozart. Strauss could write superbly for the piano—in the *Burleske,* in *Le Bourgeois Gentilhomme,* and particularly in the later songs, such as the *Ophelia Lieder,* which I recorded with Elisabeth Schwarzkopf. His piano writing is devoid of any ostentation, any exhibitionism or fake virtuosity. But he didn't choose to do much work in that genre.

That is the great pity—this gap in the piano repertoire. It was an orchestral period, and the piano was little more than a backup, a poor man's orchestra, a substitute, "first draft" kind of instrument.

T.P.: The only piano piece by Strauss that comes easily to my mind is that little "Träumerei" that used to be included in those Theodore Presser-type "Great Musical Masterpieces for Piano" collections that were so prevalent at the turn of the century.

G.G.: I'll bet that's from Op. 9, which I haven't played yet. I've played Op. 3, which consists of sturdy little pieces in the intermezzo style. None of the Op. 3 pieces have names, but all those in Op. 9 do. They're generally weaker pieces than those in Op. 3.

T.P.: My vision of Strauss is an unconventional one. Although he is often thought of as the late romantic par excellence, my favorite Strauss pieces are those from his old age, from his last period. I love the serene, nostalgic and ultimately classical purity of such works as *Daphne, Capriccio,* and *Metamorphosen.*

G.G.: Do you know the writer Jonathan Cott? A very interesting man, and a friend of mine. We've actually never met; our relationship is . . . terribly telephonic. Jonathan is a devoted, *fanatic* Straussian of the most lyrical order, and he speaks with the same reverence and enthusiasm that you do for works like *Metamorphosen, Capriccio* and the Oboe Concerto.

It's interesting: When I made a documentary about Strauss last year, I got a strong response for the last pieces from a number of the younger people I talked with ... Jonathan Cott and the composer Stanley Silverman, for example. Silverman has considerable reservations about Strauss as an opera man but, again, loves the late works. It's the elder statesmen—like Norman Del Mar, who wrote the three-volume study of Strauss—who don't think so highly of the last pieces; but then you have a young man like Jonathan going on in an ecstatic way. Extraordinary—quite the reverse of the generation gap one would expect.

T.P.: There *is* a decline in Strauss' middle period.

G.G.: Oh, no question. I've never been able to take a work like *Ariadne* seriously—in fact, I'm not fond of *Der Rosenkavalier.* But even a work like the *Alpine Symphony* ... now this is a work which has had a very bad press all its life, but there are *moments* in that piece—even though, yes, the coda *does* go on forever, and no, he doesn't seem to know how to get off that pedal point at the end [*laughs*]—but there are those moments—indeed, great long swathes—that put to shame even the best of the early tone poems. It doesn't hold together structurally in the way that something like *Till Eulenspiegel* does, but there is a seriousness of intent that simply wasn't there in the early years. And then pieces like *Capriccio!* I don't know *Daphne* that well; now that you mention it, I will have to study it.

T.P.: It's gorgeous. You can pass up the opera *Friedenstag,* however.

G.G.: Yes, I have a score of *that* one! [*laughs*] You know, Strauss was a much more abstract thinker than most people give him credit for, and the only romantic composer after Mendelssohn who never violated the integrity of what I might call the *inferential* bass of the voice-leading components in the structure of the music. (Some people would put in a claim for Brahms on that score, but he does slip up occasionally, and the rest of the time he's so *bloody* self-righteous about *not* slipping up.) *Metamorphosen* is my favorite Strauss piece, because in it he has finally come to terms with the abstract nature of his own gift. In a way, it's Strauss' *Art of the Fugue.* It's an *asexual* work, if you like, a work that has no gender. It could belong to the organ, or to the human voice, just as easily as to the twenty-three solo strings for which it was written. But, anyway, I wandered off the point, because I started out to say that it was a great shame that Richard Strauss did not write more for the piano. But I can tell you right now that I'm not going to help him out by transcribing *Metamorphosen*, because I haven't got that many fingers!

T.P.: Sibelius is also considered a late romantic, but once you get past the first couple of symphonies, there are few more austere and classical composers. You have recorded some of the piano music, which is all but unknown today.

G.G.: Yes, if you count little pieces within opus numbers— titles like "Träumerei" or "To a Fir Tree" [*laughs*] or whatever—Sibelius wrote something like a hundred and seventeen pieces for the piano. Most of them are completely insignificant, but I am fascinated by the three sonatines I recorded. They have the same spartan concision, bordering on the stingy, that is found in his symphonies, but their idiom is almost neoclassical. Quite extraordinary, considering these sonatines pre- date World War I, yet they contain an anticipation of the postwar zeitgeist. But of course they are not mas- terpieces; nothing Sibelius wrote for the piano really was. He was mainly interested in the orchestra. I *do* admire the fact that when he does write for the piano, he doesn't attempt to make it into a surrogate orchestra. It is always definitely piano writing.

T.P.: I can understand how the Nordic music of Sibelius must appeal to you, for your interest in the far north is well known. You have made a radio docudrama en- titled "The Idea of North," and I seem to remember that you once said something to the effect that it was difficult to go far north without becoming a philosopher.

G.G.: What I actually said was that most people I have met who actually did immerse themselves in the north seemed to end up, in whatever disorganized fashion, *being* philosophers. These people I met were govern- ment officials, university professors, and so on—people who had been very much exposed to a kind of unifying atmosphere. None of them were born in the north; they all *chose* to live there, for one reason or another. What-

ever their motive in moving north may have been—
and it varied from person to person—each individual
seemed to go through a particular process which greatly
altered his life.

At first, most of these people resisted the change;
they reached out, contacted friends, made sure their
subscription to *The New Yorker* was intact, and so on.
But after a while they usually reached a point when
they said to themselves: "No, that's *not* what I came
up here to do."

In general, I found that the characters who had
stuck it out long enough and removed themselves from
the sense of curiosity about what their colleagues were
thinking, or how the world reacted to what they had
done, developed in an extraordinary way and under-
went an extreme metamorphosis.

But I think that this can be true of anybody who
chooses to live in an isolated way—even in the heart
of Manhattan. I don't think the actual latitudinal factor
is important at all. I chose "north" as a handy metaphor.
It may be that the north is sometimes capable of providing a helping hand in getting people out of a situation
they couldn't pry themselves loose from otherwise; it
may be that looking at endless flowers on the tundra
during the two procreative weeks in July is inspiring,
but I don't think that the latitude is what made these
people philosophers—if indeed that is what they became. No, it was this sense of saying, "I don't really
care what my colleagues back at the University of Fill-
in-the-Blank or at the department of external affairs
think about this solitude, for *I* am going to do it, and
I am going to discover something!"

T.P.: A purification process.

G.G.: Yes. This process could have occurred even had these
people simply locked themselves in their closets...
although that might have been rather less attractive
visually.

T.P.: So you really mean the disembodied "idea" of north.

G.G.: Precisely.

T.P.: In your docudramas you often use a technique where
three or more voices are all talking at the same time,
making it very difficult to zero in on any single sentence

or idea. You have referred to this as "contrapuntal radio."

G.G.: Yes. I don't honestly believe that it is essential in radio that every word is heard. One emphasizes just enough key words in the ... countersubject sentences, if you will, so that the audience knows that voice is still happening, but it still allows them to zero in on the primary voice or voices and to treat the others as a sort of basso continuo.

We come from a long and splendid tradition of radio, but it has always been a tradition that was very, very linear. One person spoke, then the next person spoke, and occasionally they interrupted one another with an "and" or a "but." Two people never spoke together; that made no sense. I grew up in that particular tradition and enjoyed its products hugely. Nevertheless, I always felt that there was a musical dimension in the spoken word which was being totally ignored.

I coined the term "contrapuntal radio" to respond to certain criticism. When "The Idea of North" first came out in 1967, the fashionable word was "aleatory," and some critics were quick to apply this term to my work. *Nothing* could have been further from the truth, and to counter this impression, I began to speak of "contrapuntal radio," implying a highly organized discipline—not necessarily leading to a fugue in every incident, but in which every voice leads its own rather splendid life and adheres to certain parameters of harmonic discipline. I kept a very close ear as to how the voices came together and in what manner they splashed off each other, both in the actual sound and in the meaning of what was being said.

Now I am drafting an idea that I don't really expect to get to work on for a year or so, but at that point I intend to do a radio equivalent of Tallis' sixty-four-voice motet [*laughs*]—but I don't intend to say anything more about that, as it will probably jinx the whole project if I do!

T.P.: You have also worked with some of the same ideas in television.

G.G.: Yes, I've written a television script on the fugue, part of a series of five programs on Bach that I am doing for a German company. I've been having a really hard time with this project, because the rough guidelines are for forty minutes of music and only twenty minutes of talk. It is an absolutely impossible task to try to deliver any important thoughts on the nature of a fugue in twenty minutes.

There is nothing aleatoric about my television work, either. In the film, there is a discussion between myself and the director which will appear to be spontaneous. In reality, it will be the product of months of hard work, concise scripting and rehearsal.

T.P.: Turning back to your piano recordings, I'd like to talk about your oft-quoted statement to the effect that the only excuse for recording a work is to do it differently.

G.G.: That's true, but I've always meant to immediately interject that *if,* however, that difference has nothing of validity to recommend it musically or organically, then better not record the work at all.

I am not without stain in this regard, because there are works that I have recorded simply for the sake of completeness that I had no convictions about whatsoever.

T.P.: Would this include some of Mozart's piano music? Your performances of some of the sonatas strike me as possibly your least successful records.

G.G.: Yes, a couple of the later Mozart sonatas. The early works I love, the middle ones I love, the later sonatas I do *not* like; I find them intolerable, loaded with quasitheatrical conceit, and I can certainly say that I went about recording a piece like the Sonata in B-flat major, K. 570, with no conviction whatsoever. The honest thing to do would have been to skip those works entirely, but the cycle had to be completed.

T.P.: You're not very enthusiastic about much of Beethoven's work, either.

G.G.: I have very ambivalent feelings about Beethoven. I'm absolutely at a loss for any reasonable explanation as to why his best-known works—the Fifth Symphony, the Violin Concerto, the "Emperor," the "Waldstein"—ever became popular, much less as to why they have retained their appeal. Almost every criterion that I expect to encounter in great music—harmonic and rhythmic variety, contrapuntal invention—is almost entirely absent in these pieces. In his middle period—the period which produced those works—Beethoven offered us the supreme historical example of a composer on an ego trip, a composer absolutely confident that whatever he did was justified simply because he did it! I don't know any other way to explain the predominance of those empty, banal, belligerent gestures that serve as his themes in that middle period. The later years are another story—my favorite Beethoven symphony is the Eighth, my favorite movement in all of his sonatas the opening of Op. 101, and, for me, the "Grosse Fuge" is not only the greatest work Beethoven ever wrote but just about the most astonishing piece in musical literature. But even the late works are remarkably inconsistent—for instance, I don't think that the remainder of Op. 101 has much to do with the extraordinary first movement, except for that quotation right before the finale.

All in all, I'd have to say that Beethoven's most consistently excellent works are those from his early period, before his hearing started to go—let's face it, that *did* affect his later work—and before his ego took complete command. Almost all of those early piano works are immaculately balanced—top to bottom, register to register. In these pieces, Beethoven's senses of structure, fantasy, variety, thematic continuity, harmonic propulsion and contrapuntal discipline were absolutely, *miraculously* in alignment. I'm talking about the Sonatas Op. 26 and 28 and the variations like that marvelous set in Op. 34. These works have such a sense of peace, such a wonderful, pastoral radiance, and every texture is as carefully worked out as it would be in a string quartet. What I'm going to say now may surprise you—musicians are supposed to have more sophisticated tastes than this—but I think that one of Beethoven's real masterpieces is the "Moonlight" Sonata.

But even in these early years, I have to tell you that Mr. Beethoven and I do not see eye to eye on what constitutes good music. About 1801, Beethoven wrote a letter in which he stated that his best piano sonata to date was Op. 22. And much as I love the early sonatas—and I really *do* love them—there is one dud in the batch . . . and that is Op. 22.

T.P.: Do you think it is time for a return to epic forms? Many artists seem to believe this is what will be occurring in the next decade.

G.G.: I try to avoid thinking in such generalizations about musical/artistic trends. If I said, "Yes, it is time for the return of the epic," that would imply that there was some point in the past when it was *not* right to produce one. And I don't think that that is necessarily so.

Let's look at the year 1913—no, no, 1912, even better. You have Arnold Schoenberg writing *Pierrot Lunaire;* Webern is working on the short pieces that immediately follow his string quartet miniatures, and Berg is composing the *Altenberg Lieder*. If the world stopped at this point, a historian would have to say, "The Age of the Epic is over; we are now in an age of fragmentation and the breakdown of the idea of the great long-breathed line of continuity in music." I simply *cannot* believe that this would be an adequate summation of the year 1912—even though many music historians would describe its prevailing tendencies in this way. But at the same time Jean Sibelius was working on the first draft of his Fifth Symphony, which is certainly closer to epic than fragment! I don't think I need go any further than to point out the complete underlying absurdity of such generalizations.

I find it very disturbing to contemplate the lemminglike tendencies that artists in general assume— you know: The *anti*hero is in this year, the hero will be back next year. It shouldn't matter; one should be free of all that.

T.P.: Along the same lines, what would you say are the important issues confronting a composer in the 1980s?

G.G.: *Well* . . . I don't really know. I am unable to react to a situation in which a zeitgeist-compelled tendency suggests that a particular motivation is adequate or appropriate for more than one individual at a time. I would like to see a world where nobody cared what anybody else was doing—in which the entire groupthink "You hold a C-major chord for thirty minutes, I'll hold it for thirty-one" syndrome utterly disappeared. This is not entirely a contemporary problem—twenty years ago, it merely manifested itself in a different way.

For this reason, I don't have an axe to grind. I can't say, "I would like to see the reaffirmation of the tonal system in all its original glory," or "I would like to see a return to pure Babbitt serialism, circa 1959." What I *would* like to see is a situation in which the particular pressures and polarizations that those systems have engendered among their adherents and their opponents just didn't exist.

I would think that the New York music scene would be a terribly difficult thing to be involved with unless one simply lived there and was quite specifically *not* a part of it. I find it very depressing to hear about situations in which this very competitive/imitative notion of what is au courant rules creativity. I can't think of anything *less* important.

One of the things I find most moving about the final Contrapunctus in *The Art of the Fugue* is that Bach was writing this music against *every possible tendency* of the time. He had renounced the kinds of modulatory patterns that he himself had used successfully six or seven years earlier in the "Goldberg" Variations and in Book 2 of *The Well-Tempered Clavier* and was writing in a lighter, less clearly defined early-Baroque/late-Renaissance manner. It was as though he was saying to the world, "I don't *care* anymore; there are no more Italian Concertos in me; *this* is what I'm about!"

T.P.: One final question, Glenn. If a record store flew off the planet into space, and our music was picked up by alien creatures who knew nothing of the circumstances of its composition, or what the pieces were meant to represent, or what the composer's reputation was, what pieces would be taken to heart by this alien community? In this context-free situation, what would make their top ten?

G.G.: [*laughs*] Once again, I don't know quite how to answer that! But I will say that one composer who wouldn't make it—except for the last pieces and a few of the first ones—is Beethoven. He is one composer whose reputation is based entirely on gossip. The "Grosse Fuge" would make it, the early piano sonatas, maybe the Op. 18 quartets, but I don't think that there is room in space for the Fifth Symphony. Not at all.

Piano Quarterly (1981)

15 *MIECZYSLAW HORSZOWSKI*

The tiny, fragile old man stepped carefully from the wings, and, for a moment, we caught our collective breath. *Would he be able to make it to the piano?* It seemed doubtful: One suppressed an urge to jump the stage and offer assistance. But Mieczyslaw Horszowski knew what he was doing. He walked slowly, deliberately, with just the hint of a wobble, toward his destination, and acknowledged the applause with a beatific smile. Then he sat down and began to play, and our worries were at an end. For the next hour and a half, Horszowski spun out music by Bach, Mozart and Chopin—music that he has known intimately for most of a century—while the 1987 Town Hall audience listened, grateful and enraptured.

Tonight Horszowski, who turns ninety-eight in June, will play a recital at Carnegie Hall. It is probably the musical event of the season: Horszowski has been playing at Carnegie for a long time—it was the site of his American debut in 1906—but this will be his first solo concert there in more than three decades.

When writing about Mieczyslaw Horszowski, any critic must fight the temptation to turn the story into a recitation of statistics. Still, it is important to put his achievement into perspective, so bear with me a moment: Horszowski, born in Lwow, Poland in 1892, made his debut in 1901 and has been playing ever since. His is by far the longest career of any major artist: Such celebrated musical Methuselahs as Andres Segovia, Arthur Rubinstein, Leopold Stokowski and Pablo Casals don't even come close. When Horszowski was born, Richard Strauss and Jean Sibelius were in their twenties. Arnold Schoenberg was a teenager, Stravinsky was ten years old, and Brahms, Tchaikovsky and Dvorak were all still composing.

But one shouldn't make too much of these figures, for fear of turning this appearance by a still vital artist into a sort of historical freak show. Yes, Horszowski's creative longevity is impressive—indeed, un-precedented—but we do not attend his concerts just to gape at a preserved legend. I've been at those events, and they tend to be sorry affairs. ("Yes, I know she can't play anymore, but she knew Ravel and once touched Debussy!") Horszowski, on the other hand, has much to teach us.

It has been left to a new breed of critic to give Horszowski his due. The great pianistic god of the last generation was Vladimir Horowitz; other pianists were regularly held up to his image to demonstrate how far they fell from the ideal. Horszowski, thankfully, fell far indeed. Whereas Horowitz's playing was visceral, feverish, glassy and goal-oriented, Horszowski's approach is thoughtful, calm, gentle and ruminative. He is a pianist for adults, a musician's musician. I don't think that Horszowski has ever tried to "thrill" an audience; his octaves are just so-so, and he eschews athleticism for a radiant, effervescent song-fulness.

In an era that prized steel fingers and Niagaras of piano tone above all else, Horszowski was bound to be overlooked. Harold C. Schonberg dismisses him in one cool sentence ("a sound and sincere artist") in his engaging, enraging book *The Great Pianists;* Horowitz, by way of contrast, is the recipient of six adoring pages.

But, from the beginning, his fellow musicians esteemed Horszowski highly. Arturo Toscanini and Pablo Casals were among his early mentors; with Casals and violinist Alexander Schneider, Horszowski played a famous concert at the White House during the presidency of John F. Kennedy (it was later issued on Columbia Masterworks), and he returned during the Carter presidency for a Chopin recital. He is close to several young pianists, among them Peter Serkin, Murray Perahia and Richard Goode, all of whom studied with Horszowski and accord him a respect close to veneration.

He lives today in Philadelphia, in an apartment overlooking Rittenhouse Square, around the corner from the Curtis Institute of Music, where he still teaches. A bachelor for most of his life, Horszowski married a friend of many years, Bice Costa, when he was eighty-nine; she now serves as his interpreter on the rare occasions when he grants an interview. (Though his English is said to be good, he's more comfortable speaking Polish.)

Horszowski received his first piano lessons from his mother; at the age of five he astounded his listeners by playing from memory and transposing Bach inventions. He made his recital debut in Vienna at the age of eight then began the concert tours that brought him early fame as a child prodigy. When he first appeared in Manhattan, the *New York Times* compared him to Mozart.

He is the last living pupil of Theodor Leschetizky, whose other students included Paderewski and Artur Schnabel, and who helped usher in the modern approach to playing the piano. "From Leschetizky,

I learned the beautiful sound of the piano—and how to produce it,"
Horszowski once told a visitor. "I also learned from him the strongest
sense of rhythm. Without it, the music does not hold together or make
its impact."

His most celebrated concerts include appearances with Toscanini
and the NBC Symphony; a gigantic cycle of Beethoven's complete works
for piano solo, which he presented in twelve New York recitals in 1956;
a similar cycle devoted to all of Mozart's piano sonatas in 1960 and a
series of ten Mozart concertos at the Metropolitan Museum of Art in
1962.

In 1986, I heard Horszowski play Mozart's Concerto in D minor
(K.466) at the Metropolitan Museum with the Orchestra of St. Luke's.
The central Romance was played with an otherworldly delicacy, Horszowski's left hand plucking out an accompaniment that called to mind
the detached sounds of a harpsichord, while the right hand spun the
most lyrical of pianistic cantabiles. The playing seemed to emanate from
another plane of understanding, where all was temperate and serene.
And, indeed, Horszowski is virtually our last liaison with the vanished
civility of the nineteenth century.

At the conclusion of the concerto, the audience shouted itself
hoarse, calling him back again and again. He accepted the applause with
self-effacing modesty and steadfastly refused to play an encore. One
sensed that he would not insult the orchestra by putting himself in the
position of star; his approach was always that of a colleague and craftsman, at the service of Mozart and music.

Simplicity is the essence of Horszowski's art. Not the simplicity of
haste, thoughtlessness or reduction, but rather a refined essentialism
that seems tapped from the source of the music. Worldly passion is
transcended in favor of an abiding calm, as if the pianist were, to borrow
Romain Rolland's lovely phrase, "above the battle."

Horszowski has been making records for more than fifty years, but
with nothing like the steadiness one might have wished. Most of his
discs are long out of print but well worth tracking down: They include
performances of the Brahms clarinet sonatas with Reginald Kell (and,
with Walter Trampler, of the same sonatas played on viola); Beethoven's
"Diabelli" Variations and "Hammerklavier" Sonata; Chopin's first piano
concerto, and several live performances from Casals' Prades Festival.
Most recently, Nonesuch Records has set out to record Horszowski; so
far, two recital albums of surpassing beauty have been released and
should be generally available.

There can be no doubt that Horszowski's strength has dwindled
in recent years and virtuoso passages now present some difficulties;

moreover, he is now nearly blind and can no longer read from score. It matters not a bit. I can think of no pianist before the public who plays with such tenderness, intimacy and comprehension. Let us honor him while we can.

Newsday (1990)

16 CHRISTOPHER KEENE

"This is the job I always wanted," Christopher Keene said one recent afternoon in the New York City Opera general manager's office, deep within the recesses of the New York State Theater. "I wanted it from the first day I came to the City Opera twenty-one years ago as a young conductor. Matter of fact, I thought I was ready for it then. Julius Rudel, who was running the company in 1969, sensed my eagerness and very patiently said to me, 'Christopher, you can have this job someday, but right now I'm in charge and you'll just have to wait.' "

It took two decades, but, in the spring of 1989, Christopher Keene was appointed general manager of the City Opera, replacing Beverly Sills, who was herself Rudel's successor. Keene had kept himself busy in the interim; he served as the music director of the Spoleto Festival in Italy and Charleston. He led the Syracuse Symphony Orchestra from 1975 to 1984. He was the president of Artpark, a particularly innovative summer festival of the arts near Niagara Falls, from 1985 to 1989. And— most importantly, perhaps—he helped found the Long Island Philharmonic in 1979 and was its music director until this year.

Throughout most of Keene's career, however, he has maintained a connection with the City Opera; indeed, during the early 1980s, he served as music director under Sills. "I credit the City Opera with allowing me to have the wonderful career I've had," he said. "It gave me a hundred opportunities. It stayed with me when I was successful and it stayed with me when I wasn't so successful. It offered me the chance to learn and grow. And now I want to give something back. I want to bring the City Opera to that next plateau of artistic and institutional excellence. I want to make a great company even greater."

Keene has many of the attributes of a star conductor—he is regularly described as elegant, handsome and dynamic—but the first and strongest impression he makes on a visitor is that of an almost preternatural brightness. Unfailingly alert, he grasps the drift of a question and has begun to formulate an answer before it is completely out of the visitor's mouth. He speaks in complete, tapered paragraphs, with little emendation necessary. And if, in years past, Keene occasionally seemed aloof and rather testy, he has visibly relaxed and now radiates a healthy good will and confidence. And why not? He has arrived.

Tonight's performance of Mozart's *Marriage of Figaro*—the opening of the 1990 season—marks the first public manifestation of Keene's leadership, after a chaotic first year in the director's chair that was mostly taken up with administrative crises. Two months after he assumed office, several key City Opera contracts expired. When the orchestra went on strike, the remainder of the 1989 season was canceled.

The talks went on through November and weren't completely finalized until January of this year. "So basically, all of the planning we made for 1990 had to go on the back burner; because of the disruption of the season, we had to start everything all over again. The one good thing that came out of the strike is an unprecedented set of five-year contracts, which means that we won't have to deal with that sort of paralysis for a long time."

Those problems out of the way, Keene can now chart a course for the City Opera in the 90s. "My conception of the company is rather similar to the one held by Julius Rudel," he said. "Like him, I am a conductor and I approach this job as a conductor, rather than as a singer. I am dedicated to continuing the tradition of world premieres and innovative productions that made the City Opera's reputation in the early years. Beverly came to this office after a long career as a singer and her City Opera reflected that. But I'm an arts administrator by training and I've been organizing companies since I was in grade school."

Keene means this literally. Born in Berkeley in 1946 and raised in a cultured but nonmusical household, he formed his own Shakespeare troupe at the age of eight and started a chamber orchestra that played Bach's "Brandenburg" Concertos at ten. He organized one opera company in high school and another one in college, at the University of California at Berkeley, which marked the real beginning of his career. The University Opera Association, his group, staged Britten's *Rape of Lucretia* and Gottfried von Einem's *The Trial,* both relatively unfamiliar works. And then *Don Giovanni, The Medium,* and, in what can only be called a coup, the West Coast premiere of Hans Werner Henze's *Elegy for Young Lovers,* which won the young impresario a widespread and enthusiastic press.

What drove him on? "Who knows?" Keene said with a shrug. "Just a strange personality, I guess. I had a combination of talents and a compulsive, voracious personality that was never happy unless there were fifteen balls in the air. I hated being considered a child; in fact, one of the things that was great about turning forty was that it meant I was no longer that kid from California. Even by the age of seven or eight, I wanted to be taken seriously and I wasn't. I wasn't when I was fifteen or when I was twenty-five and that drove me crazy.

"But Herbert von Karajan once said that nothing that happens to a conductor in the first fifteen years is of any value because it's all luck. It's only after fifteen years on the job that good performances are to your credit. The assignment of conducting is so complex—the intellectual, personal, psychological assignment of conducting is so enormous that it take decades to begin to assimilate it."

Keene's conducting has come in for some criticism over the years. Many critics have thought him a stiff and unyielding interpreter, both at the City Opera and during his tenure on Long Island. I've found him genuinely uneven, and the occasional brusque, pedestrian performance has been compensated for by inspired readings which have too often gone unremarked. For example, he led an intricately detailed and tautly dramatic *Missa Solemnis* with the Long Island Philharmonic a few years ago and he has always displayed a feel for American music. Most recently, in Stuttgart this summer, he won raves for his performances of two operas by Philip Glass—*Satyagraha* (of which he conducted the American premiere in 1981) and *Akhnaten*.

"I have undying admiration for Keene's technical facility with new music," said John Cheek, a bass-baritone who worked with him in City Opera productions of Boito's *Mefistofele* and Jay Reise's *Rasputin.* "He was always there with a hand or a signal when you needed him. He has incredible energy and enthusiasm and he's so terribly intelligent. And, while Keene is certainly all business when he's at work, he's a very different man when he relaxes. I like him personally very much."

However, one City Opera artist who spoke only on the condition that he not be identified admits that "there is something to the complaint that Keene charges through the standard repertory. He doesn't linger much; he seems impatient to get through with it all. This trait is less pronounced in new music."

Still, Keene does not plan to go the way of James Levine who, as the artistic director of the Metropolitan Opera, also routinely gives himself the choice conducting assignments. In fact, he is in the process of cutting back on his guest conducting, picking and choosing his spots with great care. At the City Opera, Keene will be most active behind the scenes. And there has never been much controversy about his administrative abilities nor about his taste. Indeed, Keene has been a champion of new and American music from his days in Berkeley—indeed, something of a hero.

"The thing about Keene is that he has a real understanding of theater," Glass said. "I think he'll be a splendid administrator for that reason. He has a naturally authoritative manner, which he learned from conducting. And when he conducts my music, he creates a musical performance that is integral with the stage performance. It really fits it like a glove. When he conducts, he sees the stage, which a lot of other conductors don't. I think there is a very real possibility that City Opera is going to become a very exciting place in the next few years."

Comparisons to the Metropolitan Opera, the City Opera's neighbor across the plaza, are perhaps inevitable. But the two companies—for all of their white marble and Lincoln Center locations—have very different missions. A front row seat at a City Opera performance costs $45; you can get in for as little as $7. Choice orchestra locations at the Met are now $102 on weekends, $90 during the week. (There are, of course, less expensive seats available.)

But cost is not the only difference. The Met is opulent and conservative, a repository for proven masterpieces and mastersingers. It has produced only three world premieres in the past forty years and none at all since 1967 (Glass' *The Voyage* has been announced for the fall of 1992). The Met has also, for the most part, eschewed the great operas written before Mozart. The City Opera is leaner, more daring—even, on occasion, downright experimental.

"We don't see the Met as a rival," Keene said. "People often express sympathy for me because we have the Met right next door. And I tell them to cheer up because I think it's wonderful. The Met has to deal with and confront everything about opera that I despise. It is at the mercy of a few international celebrity singers, to whom it must pay exorbitant fees. It has to cater to an enormously wealthy and conservative audience which decries novelty. It has to compete with all the European government-subsidized opera houses for the services of a few conductors and singers. And it has to do operas that *nobody* is very successful at casting anymore—*Trovatore, Aida* and so on. Quick, give me your dream cast for a *Trovatore*. You see? Can't be done just now.

"Ultimately, we believe that opera is about human relationships and the expressive power of the human voice, in combination with the orchestra."

Keene believes that the City Opera will always be a cash-poor company. "But we will survive. We can draw all these wonderful artists willing to come to New York for lower fees. Once, we charged almost nothing and our prices are still low. In fact, one of the most important things we can do—morally, artistically, administratively—is to keep our ticket prices down. That is just as important for our older audience as for our younger audience. People on fixed incomes should have access to music without sacrificing half of their pensions. And we must win over the children. We already have an outreach program that reaches some 250 public schools, but eventually I want to get them all.

"Kids are great. You know, they're the best audience for modern music. They have no preconceptions; they don't whine about dissonance. Instead, they hear *sounds* and become fascinated with them and evaluate them for themselves. And they deserve to hear some opera—not just *Hansel and Gretel* but other operas as well. What about *Rigoletto?*" Keene grins. "Dead bodies in sacks—it's a natural!"

Newsday (1990)

17 MAURIZIO POLLINI

The pianist Maurizio Pollini grants interviews only slightly more often than Halley's Comet flashes through the heavens, but after his recent sold-out trio of concerts at Carnegie Hall, he agreed, after some persuasion, to sit down and endure an hour's questioning.

And so, on a clement afternoon in early spring, one day before he was scheduled to fly to Chicago for a concert and then home to Milan, Pollini invited *Newsday* to his suite in a sedate East Side hotel. There, acutely uncomfortable but doing his best to be gracious and forthcoming, he reflected upon his life and music.

"I don't give many interviews," he began, apologetically, lighting the first of what will be many unfiltered cigarettes. "Please clean up my English a bit."

In fact, Pollini's English is fine. He speaks slowly, thoughtfully, with a marked Italian accent and an occasional hesitation. He discusses his personal life reluctantly and will not speak ill of a colleague: At one point, when a negative judgment of a contemporary composer escaped him, Pollini insisted that it be stricken from the record. The impression that remained was that of a shy, intense man of enormous discipline and integrity, who has chosen to speak to the world almost entirely through music.

To add that Pollini expresses himself very well with his music would be a gross understatement. He possesses a technique that knows no difficulties, a scrupulous and unsentimental musicianship and an intelligent, questing spirit that has led him to play some of the most challenging contemporary music. Moreover, his is a unique sensibility: He follows no leaders, leads no followers.

Pollini is now forty-six years old and he has been a famous pianist for almost three decades. The son of a distinguished architect, he grew up in Milan. "I heard most of the great pianists," he recalled. "Wilhelm Backhaus, Edwin Fischer, Alfred Cortot—Cortot I heard only once and I appreciated him much more on record, I must say. Rudolf Serkin! To me he is simply the greatest interpreter of Beethoven. Nobody like him."

In 1960, when he was only eighteen years old, Pollini entered the sixth International Piano Competition in Warsaw, where he was the youngest foreign contestant and the only Italian in a field of eighty-nine. He won the competition. "Technically, he already plays better than any of us on the jury," Arthur Rubinstein, one of the judges, said at the time. Shortly thereafter, Pollini made a recording of Chopin's Piano Concerto No. 1 in E minor; it remains in print today.

"I listened to that recording recently, and liked it very much," Pollini said. "Of course, I listen to it in a distant way, because it doesn't really sound like me any more." Shortly after the Chopin disc was released, to superlative reviews, Pollini abruptly decided that he was not ready for the pressures of a career, and he canceled a projected United States tour, stopped recording and played only a few concerts in Europe. "I needed some time to think, to decide the course of my life." And so Pollini spent the next seven years studying, meditating, adding new works to his repertory and playing chess. In 1967, he resumed his career in earnest and he made his United States debut on Nov. 2, 1968, at Carnegie Hall. Again, the reviews were ecstatic.

During his youth, Pollini was an ardent socialist, but he will not discuss politics today. "Yes, I played in a few factories many years ago," he said. When asked if artists should have an active role in political life, he flashed a wry, mysterious smile. "If they do, that is fine," he said. Clearly, the subject was closed.

Pollini has recorded many different kinds of music—from the sonatas of Beethoven and Schubert through contemporary works by his countrymen Giacomo Manzoni and Luigi Nono. However, he does not agree with his colleague, the late Glenn Gould, that recordings have replaced the concert hall. "I think that concertizing is very important. But recording is fantastic. For artists, it is a testament: The performance no longer just disappears. And it brings the repertory into every corner of the world."

He prefers to record in long takes, rather than assembling a performance from several different sessions. "In the Schubert B-flat Sonata I recently recorded for Deutsche Grammophon, it was one take from beginning to end, with two little corrections here and there. Nothing more. I think you end up with the most unified performance that way. Everything holds together."

It has been said that an artist should play new music as if it were known and established, and known and established music as if it were new. "I think it is very important that the music of this century enter the repertory and become a usual part of our concert life. There shouldn't be this separation, this confrontation, between the old and the new. There is great expression in the music of Schoenberg, Berg, Webern, as much as there is in Brahms and Beethoven. But audiences, even today, find the language difficult.

"But I don't believe that one should try to change the manner of playing known repertory only to be different. I think this work survives because of its intrinsic, extraordinary vitality. Beethoven should always sound young. But this doesn't mean you have to change your interpretation to make it more modern." One should always try to find an authentic interpretation, Pollini believes, "but where it is, nobody knows!" Then he laughs.

Because of Pollini's fervent respect for a creator's intention, it is somewhat surprising to find him playing the music of Bach on a modern piano, which did not exist during the composer's lifetime.

"I had to soul search before undertaking Bach on the piano," he said. "But finally I decided that even though the piano is not the perfect instrument for Bach's music, the important thing with him was the structure, the idea, and not so much the sound or the instrument. And Bach himself made many, many transcriptions of his work, taking it from one instrument and giving it to another. And so I finally decided that the piano was all right."

Does he ever play the harpsichord or the clavichord? "For my personal pleasure, yes; in the concert hall, certainly not. The composer before Bach that I love the most is Monteverdi, and he wrote nothing at all for me. It is very frustrating."

Pollini does not read most of his criticism, and pays little mind to what he does read. "It is so hard to keep up. Sometimes I do learn something when I read the papers. Other times, I don't think it has much to do with what really went on at the concert."

He keeps his technique honed by careful practicing, every day if possible. "I don't think there is any set prescription for practicing," he said, "but after a certain amount of time at the keyboard, maybe four or five hours, it is not worthwhile going on. It might be worthwhile to look at the music—to think *into* it—but not really playing. Better to work very well for a certain time than to play many hours without the same concentration. Quality, not quantity."

Newsday (1989)

18 *STEVE REICH*

The world premiere of Steve Reich's "Music for 18 Musicians" in 1976 was heralded by stark white posters, decorated with a few measures of music, plastered on the grimy walls of SoHo. There were no advance news stories, no radio appearances, and little of the fraternal support aficionados traditionally lend to new music events. Yet Town Hall was almost full, with an audience seemingly made up of students, artists and musicians, and the 55-minute work—a succession of shimmering, ethereal aural colors propelled by a steady pulse—was followed by a stomping, cheering ovation. Tonal, repetitive, insistently rhythmic—Reich's work seemed to many a new musical idiom, unmistakably of its time, but far removed from the snarls and tangles of much twentieth-century music.

On Tuesday night, Steve Reich and Musicians will once again perform "Music for 18 Musicians," but this time they will play at Avery Fisher Hall, under the auspices of the New York Philharmonic Horizons '86 Festival. It is only one indication of how things have changed. Although Reich's records have regularly made the classical best-seller charts for some years, only lately has he been accepted by the musical establishment. His work is analyzed in conservatories, and his pieces are in the repertory of percussion and chamber groups throughout the world. He is offered more commissions than he can handle. And Reich, still viewed as a young radical in some circles, will be fifty years old in October.

"Acceptance comes with time, and I've been around a while now," he said the other day in his lower Manhattan loft, surrounded by scores, electronic keyboards, tapes and letters. Driven, loquacious, quick to laugh and to anger, the composer is an articulate proponent of his own music. "My work no longer sounds like something from another planet. A whole generation of players have grown up with it, and the fog has burned off."

Raised in New York and Westchester, Reich studied at the Juilliard School. He moved to the Bay Area in the early 60s, where he worked with Luciano Berio and Darius Milhaud at Mills College. One evening in San Francisco he discovered what would prove the beginning of his mature style when he set up two tape recorders with identical tape loops and let them run. Due to the intrinsic micro-variations in motor speeds, one machine ran slightly faster: The loops moved in and out of phase, resulting in some intriguing permutations. He applied this technique in a number of tape pieces, the best known of which are "It's Gonna Rain" and "Come Out"—his first acknowledged works. Then he transferred the phasing technique from tape to live music. In "Piano Phase" one pianist repeats a melodic pattern while another plays the same sequence at a slightly faster tempo. As the pianists move in and out of phase, inner harmonic and rhythmic relationships become apparent, giving the work a hypnotic, static quality which would become Reich's trademark.

At about this time, Reich expressed some of his ideas in an article called "Music as a Gradual Process": "Performing and listening to a gradual musical process resembles: pulling back a swing, releasing it and observing it gradually come to rest; turning over an hourglass and watching the sand slowly run through to the bottom; placing your feet in the sand by the ocean's edge and watching, feeling and listening to the waves gradually bury them."

CBS Masterworks recorded some of Reich's early music in the late 1960s, and the albums received generally positive reviews. But with the growing acclaim also came dissent, and there were those who heard nothing but irritating repetition—"stuck-record music." "Four Organs" (1970) represents Reich's aesthetic at its most austere: It consists of a single chord drawn out for twenty minutes, and inspired a near-riot when it was performed at Carnegie Hall in 1973. "Drumming" (1971) was symphonic in length (85 minutes) but scored for eight small tuned drums, three marimbas, three glockenspiels, voices, whistling and piccolo.

After "Drumming," Reich's work grew progressively more lush. "Variations for Winds, Strings and Keyboards" and "Tehillim" exist in one version for ensemble, Steve Reich and Musicians, which he formed in 1966, and another for full orchestra. "The Desert Music" (1983) is his most massive work, scored for large orchestra, soloists and chorus. Some of these works are written without repeat signs, and it is difficult to see how they can be called "minimalist," as Reich's music has been dubbed. Yet they are all recognizably descended from "It's Gonna Rain" and "Drumming."

Reich meets with some resistance from traditional Modernist composers. Elliott Carter, for one, has been vociferous and unrelenting in his denunciation. The two men were chosen to represent American music at a recent Proms Festival in London, were photographed together, but did not converse beyond surface pleasantries. "Mr. Carter's public attitude reflects poorly on him," Reich said. "Different musical styles than Mr. Carter's do exist, whether he likes it or not, and the main point is not which style you work in, but how well you compose in the style you choose."

Other Modernists have kinder words for Reich. Indeed, the Hungarian composer György Ligeti (also represented in the Horizons festival) dedicated a piece to him. "Ligeti seems to realize that American music is going to be very different from the European, although it may partake of the European heritage," Reich said. "But it's not really our job to do what Ligeti does, or what Stockhausen does, or what Boulez does. We are a different continent.

"American music is best exemplified by Gershwin, Ives and Copland," he continued. "Their work reflects American popular traditions in the same manner that Hungarian folk music is deeply imbedded in the work of Bartok. For American composers to model their work on pieces from turn-of-the-century Vienna or post-World War II Cologne is to pretend to be in another time and place, and you can't do that without suffering the consequences of being academic.

"Don't get me wrong. Berg, Schoenberg and Webern were very great composers. They gave expression to the emotional climate of their time. But for composers today to re-create the angst of *Pierrot Lunaire* in Ohio, in the back of a Burger King, is simply a joke. Vladimir Nabokov could have written a story about it."

Reich says that he has "absolutely no interest" in American atonal music. "Whereas Berg's *Altenberg Lieder*, Webern's Opus 5 Quartets and Schoenberg's Opus 11 and Opus 19 piano pieces and the Five Pieces for Orchestra (Opus 16) knock me out every time I hear them. They're *originals*, you see, and their originality shines through. But their time is over, and you just can't do Schoenberg again. Do you know how the Yankees retired the uniform numbers for DiMaggio and Ruth? Well, Schoenberg's music is as fresh as the day it was written, but his number has been retired.

"I had to learn to write 12-tone music at Juilliard and I greatly resented it," he continued. "I hope that my work won't be used as a recipe for academic music in the conservatories." Still, Reich is pleased that composers such as John Adams and Arvo Pärt have admitted a debt to his work. "They clearly have their own voices, but if my music can be useful to them in any way, then I am delighted."

He is less happy with some other appropriations. "I should be receiving royalties for the theme to 'Adam Smith's Money World,' and the whole soundtrack to the film *Risky Business*, supposedly by a group called Tangerine Dream, was an out-and-out ripoff of 'Music for 18 Musicians.' I should have sued.

"Still, if I had a dime for every trace of *The Rite of Spring* I've heard in movie soundtracks, I'd be rich," he added: "I don't think imitations will sap the power of the originals. If anything, because of familiarity with the sound, the original will come through more clearly. It will be approachable, but particularly engaging, focused and musically cogent. Ezra Pound once said that a classic is something that remains news and the best work is capable of re-creating the context of its times."

Steve Reich and Musicians, Reich's own ensemble, will be playing "Music for 18 Musicians" on Tuesday. "I want to keep writing for my ensemble," Reich said, "I think of it as a sort of garden in back of my house. I may need a particular spice from the grocer, or large quantities of food for a special occasion. But my garden has fed me, and been good to me, and I'd be foolish not to keep it going."

After the gigantic "Desert Music," Reich has again returned to working with smaller forms. He recently completed a new Sextet for the ensemble, wrote a piece called "New York Counterpoint" for the clarinetist Richard Stoltzman, and is in the middle of writing a work for the guitarist Pat Metheny. "But I'm also writing a major piece for the San Francisco Symphony, which will work with the orchestra section by section—an assembly of multitudes," he said. "I'm renting a place in Vermont this summer near a running brook. I hope it will prove inspirational."

Reich has resigned himself to being called a "minimalist." "If I'm fortunate enough to live a natural life, and history continues, I have no doubt that I will be classified as a minimalist," he said. "I don't think it's a very good description for what I am doing, but I really don't have a say. Debussy, Ravel and Koechlin were all called Impressionists, whether they liked it or not—and they didn't. Berg, Schoenberg and Webern are all called Expressionists. It's a handle, the handle on a coffee cup, something to grab on to, a way to focus a converstion. To a lot of people, I'm a minimalist. And I'll just have to live with it."

The New York Times (1986)

19 NADJA SALERNO-SONNENBERG

William McKinley, the twenty-fifth president of the United States, shot to death in 1901 by Leon Czolgosz, lives on in Canton, Ohio. This small city, located some seventy miles south of Cleveland, seems a veritable shrine to McKinley, who won his first election here as the prosecuting attorney of Stark County more than a century ago. His impassive face stares from local billboards, his name graces streets and parks and, with the possible exception of the Professional Football Hall of Fame, his memorial is the area's most celebrated tourist attraction.

One Saturday night, in the basement of Canton's McKinley Senior High School, Nadja Salerno-Sonnenberg was pacing furiously around a narrow dressing room. Upstairs, the Canton Symphony Orchestra was playing a new work by William Bolcom, under the direction of Gerhardt Zimmermann, to a packed house. Immediately thereafter, Salerno-Sonnenberg was scheduled to take the stage to play Shostakovich's Concerto for Violin and Orchestra No. 1 in A minor.

The Shostakovich—sardonic, astringent and something of a masterpiece—is not a natural crowd-pleaser, and it is probable that many in the audience would rather have heard Salerno-Sonnenberg in Tchaikovsky or Mendelssohn. But she had proven herself in Canton before and, this time around, she wanted to choose her concerto. In any event, one week earlier, when she had played the Shostakovich in Winston-Salem, N. C., no sooner had the last note died away than the audience rose, as if by command, to give her a wild, roaring ovation.

And now it was Canton's turn. "I probably need more rouge, huh?" she asked, straightening out her jacket with an irritated tug, puffing on another cigarette. In fact, she looked striking—olive-skinned, neither tall nor short, her body trim and athletic, her Mediterranean face surrounded by a thick aureole of auburn hair. But one could sense that she wanted to burst the confines of the dressing room, with its lumpy, institutional white walls that seemed to have been fashioned from cottage cheese.

"The glamorous life of a performing artist," she said with an ironic grin, as she looked over her surroundings. Salerno-Sonnenberg's voice is husky and tobacco-tempered; she speaks bluntly, with a street-smart accent she picked up as a child in Philadelphia. "Actually, this is a damned good orchestra," she continued. "Canton is right up there with Pasadena on my list of the best community groups. I'm doing two concerts here, and the Saturday crowd is the more reserved one—the audience that's tougher to please. So if it goes well tonight, we'll really knock them out tomorrow."

She has been "knocking them out" around the country for some time now, in auditoriums and on talk shows, with ensembles great and small, infusing classical music with her own do-or-die urgency. Such is her draw that the name "Salerno-Sonnenberg" on a roster will help sell out an orchestra's season, and she rarely fiddles to an empty seat.

Detractors call her playing flamboyant and theatrical. For others, she is one of the most exciting young musicians to come along in many years. The guitarist Sharon Isbin calls Salerno-Sonnenberg the "Edith Piaf of the violin." Conductor Michael Tilson Thomas phrases his approval even more succinctly. "She *is* the music," he said.

She's doing well, too: Her annual income is in the high six figures, and she has recently purchased an elegant three-bedroom apartment on the West Side of Manhattan. But she doesn't get to spend very much time there, for she plays almost one hundred concerts a year. And then there are rehearsals, recordings, television appearances...

Salerno-Sonnenberg picked up her violin, nervously played through a passage from the Shostakovich, then put down the instrument and glanced once more into the mirror. She was dressed in the same moderately high heels that she always wears onstage, tight dark pants and a strapless blouse, which she covered with a smartly tailored jacket. She made a face, goofing on the cultivated image with a tomboy's sarcasm. "That'll have to do, I guess."

Nobody came to fetch Salerno-Sonnenberg; when she heard the applause at the conclusion of the Bolcom work, she made her own way through the labyrinthine basement of McKinley High, cluttered with discarded sets from school plays gone by, and joined conductor Zimmermann in the wings. The two are old friends, and they indulged in some joking banter in the minute before they walked onstage. ("Laughter is a great release," Salerno-Sonnenberg explained later. "Even if you've played something a thousand times, you're still worried about playing it again. Ever seen Itzhak Perlman before a show? Just like Robin Williams. Hilarious!")

When Salerno-Sonnenberg took the stage, however, all was coiled intensity and she exploded into the Shostakovich. Not for her the tidy, tapered, empty proficiency that too often passes for elegance in our concert halls. She is an expressionist, and every piece she plays seems a personal battle to be won (in this sense, she is very much like one of her avowed idols, Maria Callas). Her performances may thrill or they may fizzle, but they always force one to listen anew. And, in Canton, the house came down.

"Nadja never plays it safe," her teacher, Dorothy DeLay, said recently. "She very often puts herself in danger, playing as fast as she can move, taking a phrase to a really high point. Or in a slow passage, as she draws the bow across the string, the note gets softer and softer as she sustains the tone longer and longer. You'd think it wouldn't be possible. It's breathtaking—the kind of thing you hear opera singers do."

Yet Salerno-Sonnenberg has not wanted for detractors. A good deal—probably too much—has been made of her onstage behavior. I remember a 1985 concert, in which she played Mozart's Concerto No. 3 in G (K. 216) at the 92nd Street Y, under the direction of Maxim Shostakovich, the composer's son.

She fidgeted constantly, rolled her eyes, made myriad faces, whispered loudly to the conductor during the opening tutti, tugged on her dress with the impatience of an eight-year-old in first finery, and raced offstage immediately at the conclusion of the work. She seemed frenzied, practically oblivious to the audience, as if she were wrestling with a hundred phantasms that all but had her pinned.

But those who dismiss Salerno-Sonnenberg as a high-class circus act are not listening closely. Her playing is the musical equivalent of method acting. It teases, twitches, mumbles and soars. She breaks the melodic line into twisting shards of sound, which somehow maintain their organic consistency.

Hers is not the gift of simplicity—she has a tendency to charge folkish, uncomplicated music with more import than it will bear—and the lack of discipline one finds in her stage manner occasionally spills over into her playing. But it is vital, alive and genuinely creative music-making, rather than one more dutiful bow to tradition. It is possible to dispute what Salerno-Sonnenberg does, even to dislike it. But when she succeeds, a listener is reminded of just how powerful and affecting music can be, how deeply it can penetrate, how much it can all mean.

"I don't really mind if a critic hates my playing," Salerno-Sonnenberg said over lunch at a Canton restaurant—a meal interrupted several times by well-wishers and admirers, greeted cordially. "Anyone is welcome to dislike my work, but I get angry when they call me a fake. What irritates me is the implication that I sat around with my publicity agent and decided to play the way I do to attract attention. *Do they think this is a joke?* Twirling my violin onstage is a *horrible* habit, one that I'd love to break, but I can't do it. And I've tried to play the violin without any facial expressions. I did a concert last year—I won't tell you which one it was—and I spent the whole time concentrating on keeping my face straight. And I heard a tape of the concert afterwards, and it sounded as if I'd gone to sleep. *Boring!*"

Although she remains decidedly controversial, with few critics and musicians entirely abstaining from the fray, Salerno-Sonnenberg has achieved an unusual popular success in the past few years. She has signed an extensive contract with EMI Angel, and her first recording for the label, Mendelssohn's Violin Concerto and two encore pieces, has been issued. Recordings of concertos by Tchaikovsky, Shostakovich, Brahms and Bruch are planned. And she is preparing to record an album of sonatas with her old friend, the pianist Cecile Licad.

"Cecile and I go back to the Curtis Institute together," Salerno-Sonnenberg said. "I've known her since I was twelve, and I taught her a lot of bad things—all the secret nooks and crannies at Curtis, hideouts galore. Now we're all grown up, and we live close to one another in Manhattan and I'm godmother to her child. Times change."

Salerno-Sonnenberg was born in Rome twenty-seven years ago. After her family was abandoned twice—by her father, then by her step-father—her mother took the advice of young Nadja's violin teachers and moved to the United States, so that the girl could study with the best. At the age of eight, Salerno-Sonnenberg enrolled in the Curtis Institute, one of the youngest students in the conservatory's sixty-year history.

"I had started playing when I was five," she said. "It was not my idea. Everybody else in my family played an instrument, and my mom was afraid I would get a complex or something unless I joined in. I kind of liked playing from the beginning, but never wanted to practice. I wasn't really committed to being the best violinist I could be until I was nineteen years old and already pretty good.

"Curtis was wonderful," she continued. "I think that I had a very happy childhood." When she was fourteen, she entered the Juilliard School, where she earned a reputation for an unpredictable, potentially explosive, combination of talent and irreverence.

But by the time Salerno-Sonnenberg was in her late teens, problems had developed. "It was by far the worst time of my life," she recalled, solemnly. "I had adopted a very destructive lifestyle, and I was abusing myself terribly. My grandmother had died. A love affair had broken up.

I was living in this rathole on West 72nd Street and I was snappy to everybody—my teachers, my friends. I wasn't even speaking to my family.

"I had ridden so long on my talent," she continued. "I had played with the Philadelphia Orchestra, and I'd gotten to be a pretty damned good violinist, by most standards. But I'd never really worked for anything. I had been bucking Dorothy DeLay, my teacher, for so long about changing my position that I suddenly realized that everybody else at Juilliard was playing better than I did. It was like, 'Whatever happened to Nadja?' That sort of thing. And I was just letting myself go to hell. I didn't have the courage to put one hundred percent effort into anything, to give it my all, because then I could really *fail,* and I thought it was a lot better to just be lazy and fade away than to try and then fail.

"So, anyway, I'm flunking everything at Juilliard. But I kept coming in to Miss DeLay every week, just to talk with her, you know, without my violin. She indulged me for a while, but then one week she sat me down outside her class and pointed out that it had been twelve or thirteen weeks since I had played anything for her. I said I knew that. And she spoke very quietly and said she wanted me to prepare the entire Prokofiev Violin Concerto the next week. And I said that I couldn't possibly do that, because I'd never even played through it. And Miss DeLay said 'Yes, you can do it. You can buy the music today and start practicing tonight.' And she said she wanted the whole concerto by the next week or she was kicking me out of her class.

"Now this is my surrogate mother talking! I was incredibly shaken. I called Cecile and just cried and cried. Cecile doesn't talk much, but what she said to me then really turned me around. She told me, in so many words, that I was throwing my life away and she begged me to get to work, to do something with my talent. And I listened—for once. It's hard to believe, but fifteen words from these two people changed the course of my life."

So Salerno-Sonnenberg threw herself into her music, practicing upward of ten hours a day. As she recovered and then surpassed her former mastery, she decided to enter the Walter W. Naumburg International Violin Competition.

"I knew I didn't have a chance of winning," she said, "but I thought that if I worked really hard, I might place. And so I practiced and practiced—literally thirteen hours a day—until the week of the competition, when I was so keyed up and freaked out that I couldn't even touch the violin.

"I almost got evicted from my apartment the day of the semifinals. I was frying some sausages and the place caught on fire. I grabbed my violin and my cat and ran. The super told me to get the hell out. I begged him to let me have just a few more days, and he finally did, but only those few days. I would have had to have left anyway, because I didn't have the next month's rent.

"So I got a call that night telling me I'd placed into the semifinals. And the next night they called and told me I'd placed into the finals. The finals! God! All those conflicting feelings I felt. On the one hand, I wanted to drop out; I was so frazzled that I didn't think I could ever play again—couldn't ever face a jury, an audience, anybody. But I also wanted to win."

The next morning, Salerno-Sonnenberg played music by Tchaikovsky and Ravel at Carnegie Hall, just before luncheon. "It was over. I was so glad that I tore off my gown—literally tore it off; it couldn't be repaired—and went out for lunch with some friends. And over the course of the luncheon I had four beers. And then I suddenly thought, what if there are recalls? What if they want to hear me again? I'm entered in this great competition—I'm a *finalist* in this great competition—and I'm *drunk*.

"Anyway, I went back to Carnegie and Robert Mann [the founder and first violinist of the Juilliard Quartet] came onstage and gave the usual speech—the judges were impressed, very difficult decision, and so on. Then he said that they had decided not to award a second or third prize that year. And I was sure that it was all over and I felt really low. I had finally allowed myself to believe I might have a shot for second or third place, but this killed it. Then he said that I had taken first prize, and I couldn't believe it. Total shock. My goal the week before had been to make the semifinals—to get back into the race, and start to tread water again. Instead, I won it all."

She had indeed won it all, by the unanimous decision of the judges. Salerno-Sonnenberg was the youngest first-prize winner in the history of the Naumburg competitions. "I ran to this other guy in the competition who had become a good friend of mine, and I threw my arms around him and hugged him and said 'I'm sorry, I'm sorry.' It was weird. I didn't so much feel happy for myself as I felt sorry for him. And he was great; he just smiled at me and said, 'Don't be sorry. I'm proud of you.' God, what a day!"

Since winning the Naumburg in 1981, Salerno-Sonnenberg has added several other honors to her résumé, most notably the Avery Fisher Career Grant, which she received in 1983. Fiercely articulate, with a colorful and engaging personality, she has also been the subject of several

television profiles—unusual for a classical musician—and she has become one of Johnny Carson's favorite guests on "The Tonight Show."

She has taken considerable flak for the television appearances. "Occasionally I've had orchestra managers say. 'We don't want to play with
her; she's not a serious musician.' I don't get it. Should I apologize for
the fact that I can talk? For the fact that I can be a comedian for twenty
minutes? Millions of people see those shows, and I hope that I can
provide an introduction—maybe not the ideal introduction, but an introduction all the same—to music for some of those viewers."

With celebrity have come some new difficulties, however. "I get a
lot of letters these days, and some of them are pretty weird," she said.
"I moved recently and I don't give out my address, so the letters are
filtered through my management and my public relations people. And
a lot of them are wonderful, and most of the rest of them are pretty
innocent—'I think you and I could make a pretty good team' and that
sort of stuff—but I've also been getting notes from some guy who writes
really sick, violent stuff on torn up bits of paper. And there was another
person who followed me from city to city, leaving a single black rose
in my dressing room before every show. I feel vulnerable and I don't
like it.

"I'm also tired of this schedule. I wake up sometimes and I don't
know what city I'm in. It used to be that I'd that I'd arrive in a new
town and spend the afternoon walking around, getting to know the place.
Now I get off the plane and go to the hotel room. I take care of business
and try not to tire myself out before rehearsals and the concert. There
are times when I get depressed, angry, for no real reason. Today, for
example, I have to spend the day at the hotel in Canton, doing nothing,
just resting and waiting to play. I mean, it's a *nice* hotel and all, but..."

Salerno-Sonnenberg is now booked two years in advance. But she
has arranged for a sabbatical in June 1989. "There are some technical
problems about playing the violin that I never paid attention to when I
was younger, and they're starting to bother me now. My technique isn't
really all there, and I have to work very hard to play some passages that
other violinists just breeze right through. I have seen the writing on the
wall: I really want to still be playing when I'm forty. I don't even want
to turn thirty and still have to worry about these things. I want to have
full control of the violin so I can just concentrate on making music.

"So I called my manager [Lawrence Tucker of Columbia Artists]
and said that I wanted some time off. And he said, sure, how long? I
said five months, and he didn't even blink. I was so grateful: I really
consider him a friend. But now I've got to wait more than a year before

my vacation. I've been warned about the dangers of leaving the scene. But I think I'm at the point where I can do it now, and five months sounds like such a luxury—almost half a year to sit around my apartment, practice, see my friends, cook, and have a *life* again."

She will not talk about her personal relationships. "Just write that Miss Salerno-Sonnenberg politely declined to discuss her romantic involvements," she said with a grin. "Johnny Carson always asks about that stuff, and you can't really say 'no way' on national television, so once I made up a story that I was deeply involved and terribly happy, but usually I just say something along the lines of 'I'm too busy to find the right person, Johnny.' And that's basically the truth, although I still politely decline to discuss my romantic involvements. It's nobody's business, anyway."

She is only slightly less reluctant to discuss her influences. "I never really had many heroes, you know," she said. "I looked up to David Oistrakh when I was growing up although my playing is not like his at all. At all! But I admire him; you can put on a record and hear his soul. Heifetz had a phenomenal technique, but I never liked the way he played Mozart or Bach very much. For me, he was not a complete musician, although nobody ever played the Bruch 'Scottish Fantasy' like he did. I think Callas was phenomenal—as an artist and as a woman. Such fire and intelligence. I have a videotape of some of her performances and you can't take your eyes off her."

Leonard Bernstein is the artist she most wants to work with. "I've played under him in a student orchestra, but never with him, as a soloist," she said. "I have dreams about doing the Brahms concerto with Bernstein in Carnegie Hall; I've had the same dream since I was a very little girl. He really epitomizes music for me.

"The Brahms concerto is one of my favorite pieces," she continued. "Everything is in it, from A to Z. But it's a funny piece in some ways. In that gorgeous second movement, the violin never gets the melody. First the oboe gets the theme, then the strings, but the violinist never gets it at all. It's frustrating, like having somebody you love who refuses to look at you."

Another favorite is Serge Prokofiev's first Sonata for Violin and Piano, which she has recorded and played several times in New York recitals. "It's rilly deep, man, rilly deep," she said, affecting a spacey accent. Then, serious again, "I don't know what caused Prokofiev to write such a somber piece—I really *should* know, I guess—but I've made up my own story about what is happening in the music, and it's never failed me yet.

"And I'm never more comfortable than when I'm playing Mozart," she said. "I can't really talk about music very well; it's too personal. But when I play Mozart, I think about how many years it's been since the music was written, and how many other violinists have played it since then. And I feel so honored to be part of that continuum."

The continuum is real, and Salerno-Sonnenberg has added her own distinctive contribution to it. She works hard, lives hard. In her best performances, there is always a sense that she is testing boundaries, flirting with the edge. Her persona is closer to that of a rock and roller than to the cool, calm, collected classical musician of legend (in Winston-Salem, her arrival onstage was greeted with enthusiastic war whoops from the audience).

One hopes that she will never trade impulse for routine and settle into a fastidious sameness. "If I ever bland out, it will be because of my health and nothing else," she said. "It's exhausting to play the way I do, to give your all, and I'm not so strong as I used to be. I had to cancel some dates last month, and I was not at all happy with two of my last New York appearances because I was simply worn out. I've had tendonitis, and I had bronchitis all through my Mendelssohn recording sessions—we called them the 'Chloraseptic sessions.' "

Salerno-Sonnenberg settled back into the booth and lit another cigarette. "But I could never really play without feeling it deeply. Look, this is Canton, Ohio. This performance won't make or break my career. Chances are that a lot of the people in the audience don't know the piece. I don't need to be afraid of the local critic. I could be very smug about the whole thing.

"But how can you play a work like this and not be involved— involved with one hundred percent of your body and soul? It's still Shostakovich, wherever you play it, whomever you play it for. It's great music and I owe it everything I have. And throughout this crazy business, I can take solace in the music. Whatever the critic thinks, whether or not the audience likes the piece, whether or not I'm invited back next year, I have done my best for the music I played. And that makes it all worthwhile."

Newsday (1988)

20 LEONARD SLATKIN

Outside the George R. Robinson Elementary School in Kirkwood, Missouri, a boy waited hopefully, scrutinizing every passing car. A brightly colored sign—"Welcome Leonard Slatkin!" in an array of purple, yellow and green—had been placed in front of the building and, upstairs, twenty-five children (with a few teachers and parents) sat in a classroom waiting for the guest of honor to arrive.

Four minutes late, Slatkin, the music director of the St. Louis Symphony, pulled up in his compact Mercedes-Benz and the lookout ran off to tell the news. "I've been looking forward to meeting with these kids all week," Slatkin said as he paced toward the school. Although it was the middle of a very full day—an exacting rehearsal in the morning, a meeting with a reporter from a suburban newspaper in the early afternoon and a formal dinner with the orchestra's most generous supporters yet to come—he had determined that nothing was going to keep him from this appointment.

And so, on this windy but unmistakably springlike afternoon, Slatkin drove to Kirkwood, some fifteen miles from the center of St. Louis. Dressed with casual elegance—tie, sweater and slacks—he relaxed and met Joey, Chrissy, Jenny and the rest of the Robinson school's third grade on their own terms.

"Let me tell you a little bit about what I do," he said with a grin, his back to the blackboard. "I'm what's called a conductor. That means I stand in front of a group of musicians and try to get them to play the same piece of music at the same time. It's a little bit like being a football coach. You spend a lot of time in advance, trying to get everything right so that by the time of the game, everybody knows what they are supposed to do. And then, if you get lucky, everything falls into the right place."

Informal, charming and endowed with a self-deprecating humor that is rare among conductors, Slatkin immediately won the room. "To do my job, I have to know a little bit about every instrument," he continued. "Most importantly, I have to know how they play together. And I tell them what to do, through my movements and expressions."

Slatkin then led the class through an elementary exercise in conducting, swinging his arm slowly, deliberately, with clocklike regularity. "Now follow my motion until I get . . . right . . . *here,* and then clap your hands." Pandemonium. "Well, not bad for beginners. Try it again." Slatkin worked diligently with the children until all 25 of them met (approximately) on the same beat. "There we are."

The demonstration over, Slatkin took questions, reading the names from little signs that drooped over the edges of the desks. Every inquiry received a patient answer—simplified or streamlined, perhaps, but a respectful answer all the same.

How did you become a musician? "My father was a violinist and conductor, my mother was a cellist. And when you are young, you want to do something a little like what your parents do, right? I started playing and eventually it became my life."

What did you do before you became a conductor? "Lots of things. I studied. I worked in the garment district of Los Angeles. I played piano in a bar at night. Made some good money that way. But I like conducting better."

Do you have any children? "I don't have any children but I am married and we have a dog, a spaniel called Margie. And my wife is a singer and the dog likes to sing along."

What do you call that stick you wave? "Byron." The room explodes into mirthful giggles. "No, no, no. It's called a baton. B-A-T-O-N. Cheerleaders use batons but theirs are larger and they throw them into the air. I use the baton particularly in music that is very precise and clear. It's easier for the musicians to follow than my hands."

Do you own a limousine? "Nope. Sometimes orchestras keep limos to pick up the conductors and the guest artists. But I don't want one. Who would want to pay for all that gas? Does the St. Louis Symphony own a limousine? No. We barely own our concert hall."

At length, the questions exhausted, Slatkin walked around the room and signed autographs for every child.

"Now let me ask you a couple of things," he said. "How many of you have heard a symphony concert?"

A surprising number of hands fly up immediately.

"Uh-huh. Very good. And how many of you want to be musicians when you grow up?"

Unanimity.

Slatkin smiled.

"Good for you. Stay in touch."

And he was off, for a quick nap and the long night ahead.

Powell Symphony Hall in St. Louis is one of those magnificent movie temples so much a part of the American 1920s, when money and imagination ran wild and extravagance was the order of the day. Designed by the Chicago architectural firm of Rapp & Rapp, Powell Hall features a grand foyer modeled after the Chapel of Louis XIV at Versailles, complete with crystal chandeliers and a grand marble staircase. The auditorium itself resembles nothing so much as an elaborate birthday cake, covered with stucco icing and spangled with gold leaf.

On a Friday morning, a few minutes after ten, sounds as opulent as the hall are seeping out into North Grand Boulevard and into the heart of St. Louis's depressed midtown—an area that has, with pardonable civic boosterism, been rechristened Grand Center. Such joyful noise in these surroundings is a disconcerting experience: As recently as 1980, commercial vacancy in this part of the city was greater than 60 percent, and the neighborhood retains a forlorn, postwar atmosphere, rubble mingling with renovation.

But the Middle West is nothing if not paradoxical. This bejeweled souvenir of vaudeville and the silent film, transformed into Powell Symphony Hall, is now the first permanent home of the St. Louis Symphony, which many critics and listeners would name as one of the finest ensembles in the United States.

This morning the orchestra is rehearsing a Sibelius symphony under Slatkin, the tenth music director in its 108-year history. His manner with the players is that of a firm but patient big brother. His comments are specific, authoritative but never dictatorial: Although he is unquestionably in control, it is an easy—even friendly—control.

"I use my hands like a sculptor, to mold and shape the sound I want, to clarify," Slatkin said one day after rehearsal. "I take technique for granted. In this day and age, conductors, like singers and pianists, are simply expected to be technically proficient. But technique can only be a beginning. After all, many of the greatest conductors—Toscanini and Furtwängler are perfect examples—were only competent technicians. An understanding and ability to communicate the music to the musicians is what matters."

Slatkin is the antithesis of a podium glamour boy. His gestures are passionate, but devoid of histrionic excess; during particularly rapturous moments, his eyes will close momentarily and on occasion he will allow himself one, quick choreographed leap into the air, but he is playing for the musicians, rather than any prospective audience. His stick technique is transparently clear—the expression concisely conveyed, the beat unmistakable. Orchestra and conductor seem fellow pilgrims on a quest for the Sibelius Second.

The orchestra responds reflexively to Slatkin's demands—the tympani right on time for a clap of Sibelian thunder, the flutes wild as Northern birds. A perfectly contoured crescendo sweeps from near inaudibility to an explosion of sound that fills the entire hall. As the last chord dies away, so clean and unanimous that it seems to have been produced by a seraphic organ, the musicians break into startled laughter. Are they *really* playing this well?

They are indeed. Slatkin, red in the face and drenched with sweat, looks out at his spent forces and grins. "O.K.," he says, with calculated understatement. "I think I can live with that."

Under Slatkin's direction, the St. Louis Symphony has developed to a point that it is regularly compared with the leading ensembles from New York, Boston, Philadelphia, Cleveland and Chicago. And Slatkin has become the first native-born American to conduct all of these orchestras—the so-called "Big Five"—in the course of a single year. Which is quite an accomplishment, for our performing arts organizations suffer from a national inferiority complex that has led them to look, for the most part, toward European leadership.

"Concerts wake me up," Slatkin said one evening as he ordered a full-course dinner in a St. Louis restaurant. "It takes me a couple of hours to wind down. I lose between two and three pounds during a concert, depending on the music and the shirt I wear." He had just led an ambitious program consisting of "Decoration Day" by Charles Ives, the Hindemith Violin Concerto and the Sibelius work. After acknowledging the applause, he dashed back to his dressing room and greeted some well-wishers. And then it was off to Tony's, a favorite spot, for the ritual late meal.

"We were 98 percent sold out tonight," he continued. "I'll get the exact ticket count tomorrow. We're still building a Friday night audience; our Saturday and Sunday performances inevitably sell out."

It is a five-minute drive from the center of town to the Central West End of St. Louis, where Slatkin lives in a spacious old house. Penguin images are everywhere—glass penguins, sculptured penguins, toy penguins, hundreds of penguins scattered throughout the house. It is the visual equivalent of a motto or, perhaps, a *leitmotiv*.

"All my penguins are gifts," said Slatkin. "A friend of mine started me on this obsession. She gave me my first penguin, making note of the resemblance to a conductor in tails. The one thing I still want is a picture of myself with a penguin, which the St. Louis Zoo cannot allow. It turns out that penguins are terribly susceptible to human diseases; they've built up very few immunities in Antarctica."

Slatkin has deep roots in St. Louis, professional as well as personal. It is difficult to think of any American conductor since Eugene Ormandy who has been so closely identified with the city in which he works. Today conductors tend to be cosmopolitan animals, jetting here and there while maintaining homes in a musical capital such as London or New York. Although Slatkin does a lot of touring, he has kept a residence in St. Louis for almost twenty years, since he was first engaged as Walter Susskind's assistant conductor. It has been suggested that Slatkin has shaken hands with every man, woman and child in the metropolitan area. Slatkin demurs; he says there must be some that he's missed. But his involvement with the city is legendary. And the affection is returned. The local feeling about the St. Louis Symphony is a little like the local feeling about the St. Louis Cardinals: It is one thing the city owns that everybody—inside and out—agrees is first class.

Slatkin is much in evidence in St. Louis. He attends as many home baseball games as he can (he tried, without success, to talk the Symphony's guest artist Marilyn Horne into singing "The Star Spangled Banner" at a 1989 Cardinals game). He knows the back roads, where to get the best Italian food downtown and the best frozen custard in the suburbs.

His father, Felix Slatkin, grew up in St. Louis and was, for a time, the assistant concertmaster of the St. Louis Symphony under Vladimir Golschmann. Indeed, when Leonard first came to town, his father's violin teacher was still playing in the symphony. "If I counted all the people here who have told me that they used to live next door to my father," he said, "he would have had 2,000 next door neighbors."

In 1937, Felix Slatkin, having been refused a $5 weekly raise, quit the orchestra and moved to California, where he quickly established himself as a leading studio musician in Hollywood. There Leonard Slatkin was born on September 1, 1944.

To say that Leonard Slatkin grew up in a musical household seems an understatement: His father was then concertmaster at Twentieth Century-Fox while his mother, Eleanor Aller, was the principal cellist at Warner Brothers. In addition to his duties as a violinist, Felix Slatkin was also conducting both the Hollywood Bowl Symphony Orchestra and the Concert Arts Chamber Orchestra.

But the most durable musical activity of Slatkin's parents was undoubtedly the Hollywood String Quartet, which was founded in 1947 and continued through 1961. Because of studio commitments, the group, with Felix Slatkin on first violin, Eleanor Aller on cello as well as Paul Shure on second violin and, at different times, Paul Robyn and Alvin Dinkin on viola, rarely toured, but it made many recordings for the Capitol label. In the years since the Hollywood String Quartet's dissolution, there has grown a following of increasingly vocal admirers, many of whom consider it America's finest chamber ensemble to date and who hunt down old recordings with voracious appetite. Two multidisc reissues from EMI have proved steady sellers and more are promised.

Nobody could be happier with this reappraisal than Leonard Slatkin. "This was a West Coast group and, as such, had to deal with Eastern chauvinism throughout its lifespan," he said. "I mean, after all, how could a New Yorker possibly take something called the Hollywood String Quartet seriously? But the group had quite a reputation among musicians. I grew up with the sound of the Hollywood Quartet ringing in my ears. My first strong musical memory is of the Villa-Lobos Sixth Quartet, which my parents were rehearsing. I remember that it reminded me of big teddy bears dancing around."

Eleanor Aller, still active in the Hollywood musical community, can recall young Leonard sitting on the staircase, listening intently to recordings of the late Beethoven quartets. Visitors to the Slatkin house included celebrities as diverse as Arnold Schoenberg, Igor Stravinsky, Frank Sinatra and Danny Kaye.

It was all but inevitable, then, that Leonard would be drawn to music. He began playing the violin when he was three years old, took up piano at 12, viola at 14 and later studied composition with Mario Castelnuovo-Tedesco. He also became an avid record collector. "Leonard started spending all of his money on records when he was about seven or eight," his mother recalled. "He loved all kinds of music; one that I remember was Doris Day singing about her secret love. He drove us crazy with that one. He played it constantly."

His parents' schedule did not make for a particularly intimate family life. "My brother and I were raised by a succession of housekeepers, some of them pretty bizarre," Slatkin remembered. "One, after cooking us a fabulous dinner, packed her bags and fled when the quartet began to rehearse. It turned out that she hated music, and my parents had forgotten to tell her about their profession.

"The family always maintained a kind of professional relationship. I still don't talk much with my mother about professional matters. Because we all did the same thing, we were very sensitive to each other's criticism and kept things pretty much on a musician-to-musician basis."

Slatkin's younger brother, an accomplished cellist who performs under the name of Frederick Zlotkin, concurs. "I remember a party for my father in which every member of the household wrote a variation on 'Happy Birthday.' We were all so determined to outdo one another that we all wrote variations so difficult that nobody could play them."

Slatkin remembers himself as an introverted child. "I liked isolation then and I still do today. There is a difference between being alone and being lonely. It's true that I had few close friends. But I had my scores and recordings, which allowed great composers to speak with me. To this date, I feel in close communication with my composers; when I am preparing a piece, something, somehow, tells me when I am on the right track."

In 1962, Slatkin left Los Angeles and went to study at Indiana University. He lasted only one semester; the university had compulsory R.O.T.C. training at the time and Slatkin refused to participate. "I am a lifelong pacifist and I didn't want to have anything to do with the military." He returned to California to study English at Los Angeles City College. "I got back in January 1963. The next month my father suddenly died of a heart attack. He was an alcoholic and I am convinced that his heavy drinking contributed to his death. My father drank to cover up his insecurities, I think, and since we have the same temperament, I am very careful with alcohol. I'll have a little wine with dinner but that's it.

"I'm very sorry now that I didn't know my father better. It's strange, you know; after he died, I couldn't remember a thing about him. It was as if 19 years of my life had ceased to exist. My memories were literally erased. So I went through a period of soul searching, spending a year and a half and batches of money, going through my life. And then I started to get curious, and I asked my mother all sorts of questions about my father—what he was like, how he felt about this or that. As a result, I feel much closer to him now that I ever did when he was alive.

"I do not think that I ever would have become a conductor if my father had lived. I could never have challenged him in that way. In 1983, when I recorded the soundtrack for a film called *Unfaithfully Yours* with Dudley Moore, we did the taping on the same soundstage where my father had worked for so many years and where I had visited him so often. I almost couldn't start: The place was so familiar and the circumstances so strange that I almost broke down."

Slatkin describes the years immediately after his father's death as a "period of floating." He began to do some conducting for a community orchestra, almost by accident, but felt lost and confused. Then, in the summer of 1964, he went to the Aspen Music Festival to work with Susskind. This led to acceptance at the Juilliard School, where he studied with Jean Morel from 1964 to 1968.

"I liked Juilliard," he says now. "There was a lot of pressure but there weren't as many conductors as there were violinists and pianists and so you didn't find the same sort of cutthroat atmosphere that other graduates complain of. I did a lot of work but there were some fun times, too. I used to shoot pool with Pinky Zukerman.

"By 1966, I was the assistant conductor of something called the Youth Symphony Orchestra of New York. I don't remember the exact date of my first concert but I will never forget stepping out on to the stage of Carnegie Hall to lead the 'New England Triptych' by William Schuman."

When Susskind accepted the job of music director in St. Louis, he invited Slatkin, 23 years old and fresh from Juilliard, to become his assistant. There, Slatkin conducted the Sunday afternoon series of concerts and occasionally appeared as a piano soloist. He progressed to associate, associate principal, and principal guest conductor before assuming the position of music director in 1979, replacing Jerzy Semkow.

"One of the reasons I have such an easy rapport with the orchestra is because I really grew up, musically speaking, here in St. Louis," Slatkin said. "I started off as the assistant conductor and part of my job was to calm down the players when they were upset about the music director or a visiting artist. I was very chummy with everybody, went to dinner and to parties with them. And then my role changed and became awkward because, as music director, I had to consciously break some important social ties.

"But I try to keep things relaxed. At my first rehearsal as music director, our principal bass player addressed me as 'Maestro.' I said, 'Wait a minute, Henry, let me hear you say that again.' He repeated himself. And I said, 'No, I think Leonard will do fine.' It's true that some of the newer members call me Mr. Slatkin but 'Maestro' seems a little formal.

"I am delighted with the sound we have achieved here and I try to take a little bit of St. Louis with me wherever I conduct. But you can't do that with every orchestra you work with; if you're just visiting, you don't want to rearrange the furniture too much.

"A great deal of the conductor's job is done before setting foot on the podium. I try to come in with a clear conception of the way a piece should be played and then I make adjustments for the personalities of the different orchestras I work with.

"Above all, I strive for a fat, luxurious string sound. I try to get some kind of unanimity from the winds about when to breathe, and to convey that conception to the strings. I don't like blaring horns. The brass should be kept somewhat in abeyance; for me the orchestra revolves around the string section."

Slatkin singles out Fritz Reiner and Carlo Maria Giulini as two conductors who made a strong early impression. "In addition, any American conductor must hold Leonard Bernstein as a role model. Whether you agreed with everything he did or not, he single-handedly put the American conductor on the map. He fought all the battles for the rest of us, and attained a respectability the American conductor had never known before.

"I also admired Eugene Ormandy a great deal. I can truthfully say that I never heard a bad concert from Ormandy. He may not have been fantastically inspired every time he set foot on stage but he was a craftsman, always dependable. And he was a great accompanist. That's very important. Somebody can lead the most thrilling Mahler Ninth in the world but if he can't accompany a Chopin piano concerto, I don't think very much of him as a conductor.

"A music director cannot do everything well," Slatkin continued. "I know my limitations. For instance, I'm still learning about opera because this is basically a new interest. Symphonic conductors work in the abstract while operatic conductors must fit music to words, sentences, a story line. I grew up with a string quartet in residence at my house and when you place most of the operatic repertory side by side with the Beethoven, Mozart and Bartok quartets, it seems pretty feeble. There simply isn't the spiritual intensity. And, with the exception of Mozart, there is not one composer who was equally at home in the chamber and operatic repertory, so I was not exposed to many operas at home.

"Even though I conducted the American premiere of that little symphony Mozart wrote when he was nine years old—who turns down an opportunity like that?—I still have something of a misconception of Mozartean style. I grew up on romanticized Mozart and still carry that sound in my head. But I'm learning. There is a line I must learn to walk. The problem is to achieve the crystalline sound that one gets in a small-scaled, musicologically correct Mozart performance without making him sound prissy. For, after all, this is intensely masculine music and, in some ways, Bruno Walter's performances, romantic or not, came closer to the heart of the matter than many more authentic performances.

"A great deal of my interest is in twentieth-century music but only specific works. Mahler is overrated: Some of his symphonies are magnificent but they cannot stand with the Beethoven nine, the Brahms four or even the Sibelius seven. Some of his music is inspired, some is heartfelt, but other works are superficial, even bombastic. I don't like much 12-tone music—none of the later Schoenberg, not much Webern. And, although I know there are people out there—even some genuinely

musical people—who sincerely admire Elliott Carter, I don't hear much
in his work at all. It's just a series of mathematical gestures, piled on
with needless complexity."

Carter and Slatkin have had some public differences. Before a 1983
Chicago Symphony performance of Carter's "Symphony of Three Orchestras," the composer walked out, objecting to what he called the
"giocoso" nature of Slatkin's informal introductory remarks. Slatkin
shrugs off the incident. "I always like to talk about a difficult piece
before I conduct it," he said. "It makes things more interesting for the
audience. I meant no disrespect to Mr. Carter. Simply because I don't
like a particular piece of music doesn't mean I can't lead a performance.
I even recorded the Pachelbel canon.

"On the other hand, I still don't like Mr. Carter's symphony. But
there is a lot of fine new American music that manages to speak in a
contemporary idiom while maintaining a certain accessiblity. I have admired works by David Del Tredici, Joseph Schwantner, Steve Reich,
Dominick Argento, John Adams. We have commissioned works from
several of these composers. And, although I respect Oliver Knussen,
Heinz Karl Gruber and some other European composers, I would rather
champion the music of my native land than go looking across the ocean
for cultural models."

Off stage, Slatkin's main interests are films and sports. "If it's a
movie, I'll watch it, even if it means 15 and a half hours of *Berlin-
Alexanderplatz*. And, although I don't read as much as I used to, when
I'm on vacation, me and Stephen King are the best of friends."

Slatkin is an avid sports fan and he has been known to announce
St. Louis Cardinals baseball scores during concerts. "I'm not insensitive
to the fact that there are some people in my audience who don't want
to be there, and would much rather be at home watching the game. At
times, I'd like to join them myself.

"But most of the time, I want to be involved in music. Even when
I am reading or driving my car, I am listening to music. I listen to
everything—whether the symphonic repertory or Chick Corea or Miles
Davis.

"My main interest is classical music, of course, and I want to do
something for some composers who have not been given the respect
they deserve. Rachmaninoff, Vaughan Williams, Sibelius, Nielsen and
the later Richard Strauss fascinate me. They all have one thing in
common: They faced the twentieth century squarely and decided that
they couldn't enter it. They were all uncomfortable—philosophically,
harmonically and even rhythmically—with the musical developments of
this century, with both the Schoenberg line and the Stravinsky line. But

I don't find that a fault, for all of these composers had unique voices, unique personalities."

A great orchestra must have a unique personality as well. The St. Louis Symphony is the second oldest in America, preceded only by the New York Philharmonic. It was founded in 1880 by Joseph Otten, a German organist, as the St. Louis Choral Society, with the expressed purpose of satisfying the insatiable late nineteenth-century passion for cantatas and oratorios, utilizing both amateur and professional musicians. The St. Louis Choral-Symphony Society was formally incorporated by 1893. By 1910, there was a regular 20-week season and, in 1923, the orchestra made its first recordings.

It was during the tenure of Golschmann, who was appointed music director in 1931 and remained in St. Louis until 1958, that the orchestra became firmly established as one of the best of our second-tier orchestras, and an ensemble capable of some exciting musical surprises. After the orchestra's initial New York appearance, at Carnegie Hall in 1950, Virgil Thomson called the St. Louis "an accomplished and well-trained group, sensitive as to nuance and rich of tone."

Today, the St. Louis Symphony has 101 full-time members and a 52-week annual season. It also has a $350,000 deficit, which worries Slatkin considerably. "We have put a frank appeal for funds into our program guide," he said. "We've been here for more than one hundred years and I can't see us folding anytime soon but who knows what will happen eventually? I'll really start to worry when we begin to lose good players for economic reasons. This hasn't happened yet, but we've had some great strikes—most recently during our 100th anniversary season. I can sympathize with the players, however. In 1984, we had a minimum base pay of only a little over $500 a week while up in Chicago they were making about a grand. I don't blame the orchestra for its annoyance."

In 1989, Slatkin renewed his contract with the St. Louis Symphony, ensuring his presence there through the 1992–93 season. He had been widely discussed as a possible successor to Zubin Mehta at the New York Philharmonic, when Mehta leaves in 1991, but the St. Louis contract makes his candidacy less than likely, unless the Philharmonic is willing to be very, very patient or Slatkin is willing to be very, very busy. Such an appointment is not entirely out of the question: The Philharmonic has waited several years before choosing a music director in the past—dividing up the duties as necessary—and it is not unheard of for conductors to be associated with more than one orchestra at a time. (For example, Gerard Schwarz was recently the leader of ensembles in New York, Los Angeles and Seattle—simultaneously.)

"A lot of things contributed to my decision to renew my contract but the bottom line is that things are going very well indeed in St. Louis," he said. "Leaving it would be like leaving a good marriage and you don't just break up a marriage unless something is seriously wrong.

"This is my home," Slatkin said softly as he drove through the city. "No matter where I travel, no matter how many great orchestras I conduct, I love to come back to this orchestra because it is mine. Whether people agree or disagree with what I do, with the sound we produce, it's my furniture, my bed, and I feel comfortable in it and can take full responsibility for it."

Portions of this article first appeared in The New York Times Magazine *(1984) and* Newsday *(1989)*

21 ELLEN TAAFFE ZWILICH

"Do you remember those time-lapse photography films they used to show in high school biology classes?" Ellen Taaffe Zwilich asks. "Years of growth were compressed into a couple of minutes. First you saw a root, then a sprout, then suddenly the tree began to grow branches, reaching out in every direction. It's as if the tree were dancing. Composers grow the same way. We twist upward, while trying to keep our roots and balance."

It is lunchtime, and Mrs. Zwilich, who became the first woman to win the Pulitzer Prize for music in 1982, has shelved her pencils and music paper for a few hours. "I do most of my composing in the morning," she says. "Music's been running through my subconscious all night. So I get up, turn off the phone and become unavailable."

Zwilich is open, friendly, unpretentious and almost disconcertingly without neurosis. She is also articulate: "I think every musician understands the Pied Piper story. Music is this vast seductive force that draws you on, and you follow wherever it may lead. Don't misunderstand me; I'm no wide-eyed Romantic and I don't underestimate technique. I believe in being as conscious a composer as I can be and I do a lot of thinking before I begin a new work. But then, once I am writing, something mysterious happens. Something beyond explanation; not so much an escape as a confrontation with a deeper reality. You must be prepared, once a new work is under way, to let it take you somewhere that you've never been before."

Wide-eyed or not, Zwilich is, despite her disclaimer, something of a Romantic, at least in comparison with many twentieth-century composers. Igor Stravinsky believed that art was synonymous with technique—pure and simple. Paul Hindemith, when asked where he found inspiration, held up his pencil. In the 1950s and 1960s, it was fashionable to give compositions literalist titles like "Notebooks," "Structures" and so on. There was little talk of mystery, instinct or the subconscious.

Still, it would be simplistic to label Zwilich a neo-Romantic, for her work has a notable degree of classical poise. Yet it is more impulsive than most neo-Classicism and none of the other "isms" that make up our inadequate glossary of musical terminology quite apply. She writes in an idiosyncratic style that, without ostentation or gimmickry, is always recognizably hers. In her early works, one hears the influence of many composers—the String Quartet (1974) blends the knotty intensity of Bartok with the languorous emotionalism of Berg. In her later music, one finds a clear, logical and seemingly inevitable sense of structure—arching, charged melodic lines, aggressive rhythms and a prismatic combination of instrumental colors. Her works reflect a concision and craft that appeals to both professional musicians and the general audience. Her music is directly emotive, yet devoid of vulgarity and characterized by a taut chromaticism that stretches the limits of tonality while rarely venturing outside of them.

Mrs. Zwilich has what might be described as healthy, all-American good looks: big-boned, blue-eyed and blonde, she is the Hollywood "girl next door" grown to womanhood. She speaks softly with a husky, hybrid Florida accent. For the past 20 years, she has lived in a modern, highrise one-bedroom apartment in the Bronx. "It's been described as one part serene sanctuary, one part cluttered workspace, which seems fair to me. But it's a fifteen-minute drive to Lincoln Center, my desk overlooks the Hudson River and, on a clear day, I can see all the way to the Tappan Zee bridge." She laughs. "It's *inspirational.*"

Like many composers, Zwilich, the daughter of an airline pilot, was writing music before she formally knew how. "I used to simply make things up on the piano and play them again and again; I didn't write anything down until I was about 10. By that point I had begun studying with the neighborhood piano teacher in Miami, where I was born and brought up. It was an unhappy relationship; she made me play all these silly children's pieces and I thought my own compositions were better."

By the time she was in her teens, Zwilich was proficient on three instruments—piano, violin and trumpet. She wrote a high-school fight song, was the concertmaster of the orchestra, first trumpet in the band and a student conductor as well. She continued to compose and, by the age of 18, she was turning out full-scale orchestral works.

Zwilich attended Florida State University in Tallahassee, where she majored in composition. While there, she played in an orchestra for conducting classes given by the late Ernst von Dohnanyi, a highly respected Hungarian pianist and composer. "Dohnanyi was essentially a nineteenth-century European artist—very Old World—and I was glad to be exposed to that sensibility. Meanwhile, I was playing jazz trumpet, singing early music with the Collegium Musicum and composing. I had a ball; it was a very open kind of place. Everything I wrote got played immediately."

She received her master's degree from Florida State in 1962. After one dreary year teaching in a small town in South Carolina, she moved to New York to continue her violin studies with the legendary Ivan Galamian. She quickly established herself in the ranks of New York's freelance violinists and also spent a season working as an usher at Carnegie Hall. During this time, Zwilich also played in the violin section of the American Symphony Orchestra under the direction of Leopold Stokowski.

"Stokowski was both a classic nineteenth-century Romantic, and a visionary crusader," Zwilich remembers. "He could be maddeningly inexact—especially in the modern repertory, which he nonetheless championed. And he was something of a martinet—one of those idealists who fight selflessly for any cause that comes along, but are still capable of unthinking, autocratic cruelty to individuals. Still, you have to credit Stokowski for his belief that the conductor had a mission in life, and that part of that mission was to perform works by contemporary composers.

"I was already aware that I wanted to compose more than I wanted to play. So I kept my ears open—I was always listening to the different sounds the orchestra made, the details of the ensemble sonorities, the range of the various instruments.

"Composers need some kind of hands-on musical experience—either as conductors or performers. If you know the orchestral repertory only from studying scores and listening to finished performances and recordings, you can't really know all that's going on in the music. A score is, at best, an indication, not a final product. Playing in the orchestra allows you a first-hand experience of the subtleties that fall between score and performance." While with the American Symphony, she met and married Joseph Zwilich, a fellow violinist and a member of the Metropolitan Opera orchestra.

In 1970, Zwilich entered the Juilliard School of Music, to study with Roger Sessions and Elliott Carter. "They were immensely helpful in my development. They allowed me my independence, which is the best thing you can say about a teacher. I don't think my music sounds at all like either of them, but they influenced my thinking irrevocably."

In 1975, Zwilich became the first woman to receive a doctorate in composition from Juilliard. Pierre Boulez programmed her "Symposium for Orchestra" (1973) with the New York Philharmonic, and she began to receive critical praise and commissions. She produced a string quartet (1974), a sonata for violin and piano (1974) and the splendid Chamber Symphony in 1979.

That same year, at the age of forty, Zwilich was suddenly widowed.

While attending a performance of the Stuttgart Ballet at the Metropolitan Opera House, Joseph Zwilich had a massive heart attack and died in the auditorium where he had served for many years.

"It's still very difficult for me to listen to the Chamber Symphony," Zwilich says today. "I had begun writing it before Joe died, and when I came back to complete it, everything had changed. It was a crucible of sorts. I loved Joe very dearly, and miss him to this day, yet his death taught me nothing so much as the joy of being alive—the joy of breathing, walking, feeling well, swimming, the joy of being human. Suddenly all talk of method and style seemed trivial. I became interested in meaning. I wanted to *say* something, musically, about life and living.

"We've had to come to grips with an incredible amount of evil and pain in this century," she continues, "and you can see, hear and feel it in a lot of 20th-century art. But this agony is only one reality; we shouldn't forget beauty, joy, nobility and love—greater realities which artists must learn to express once again.

"Artists have become hung up on a faulty maxim—'Progress is our most important product.' It's a dangerous way of thinking because with continual progress can come a continual built-in obsolescence. For this reason, I think we've become righly disillusioned with progress for its own sake. Besides, if you're a sensitive listener and you listen to Bach, you don't feel that it is somehow less advanced than Mozart, who is in turn less advanced than Beethoven. That's all a hangover from 19th-century musicology: add to the language, extend the boundaries, push back the frontier. Meanwhile, there is another frontier which 20th-century composition has sometimes neglected—an internal frontier. After all, no matter what musical language you choose, every composition is a totally new creation, yet also, in one way or another, bound to the past.

"Since the 1950s, America has had an extremely diverse musical culture, yet nobody's paid much attention to it. You must remember that during the time a lot of press was being given to what was called 'academic serialism'—an inadequate name, by the way—there were other people hard at work following other visions. William Schuman and David Diamond were writing symphonies. Alan Hovhaness was writing what could be considered an early brand of minimalism. A lot of different musics have been flourishing, simultaneously."

Unlike many composers, for whom professional gossip is an integral part of the business, Zwilich refuses to discuss her colleagues—either personally or aesthetically. "I hate the bickering between different groups of composers. I'm not at all threatened by artists who make different choices than I do, and I feel supportive of many kinds of activity.

"Instead of living with the combination of pleasure and anxiety that makes up a diverse culture, some people opt for an exact set of formulas to govern their actions—politically, religiously, musically. Instead of weighing different possibilities and then making moral and aesthetic decisions, they chase absolute authority, with all choices ready made.

"Technology has given us a very precious gift—we can examine many different possibilities. A lot of composers of the past only knew their own music, and the music of their community, their era. But we can listen to Michael Jackson one day, spend the next enjoying Balinese gamelan music or follow a recording of 18th-century music played on original instruments with some new sounds for computer. This amazing range of visions—too bad there's no musical equivalent of that term— is unique and unparalleled.

"The whole idea of an avant-garde implies one single, forward-moving stream, and that's a nineteenth-century conceit. It's not only morally wrong to fight for supremacy but also a complete misunderstanding of the imperatives of our own age. Composition is about expression; it has absolutely nothing to do with who's going to end up on top in the 90s.

"Mr. Sessions once said that music was beyond emotion. That's why Mozart was able write his clarinet concerto while he was dying. Throughout musical history, you have that dichotomy—serene pieces created during miserable times, and the other way around. I think this is because composers separate, to some degree, their creativity from their day-to-day lives. Everything that happens to me adds to my understanding, but I go to another level—beyond understanding, in a way—when I write. And what I find is unpredictable.

"I'm lucky. Music comes fast and furiously to me. But even at that speed, I'm still writing much slower than my perceptions. I direct my energies entirely on one composition at a time. It's possible to copy out another work, or edit, or revise, or what have you, but I can only actually *compose* one work. In the course of a year, with any luck, I finish three or four new pieces.

"Although I only spend a few hours a day actively engaged in writing music, I think being a composer is like being a writer of any sort. There's really never a moment when you're not working. I attend concerts as a composer, think about life as a composer, listen to records as a composer, and everything adds to the music, one way or another."

The question is an obvious one, and Zwilich has answered it many times. "Why have there been so few women composers? It's simple: We were, for the most part, denied access. Still, we're finding out that there were some women who continued to compose, knowing full well they'd never hear their music. It's an incredible testimony to the creative spirit that women, despite all odds, wrote music through all those quiet centuries.

"Compare writing a poem to creating a piece of music. Once you've got those words down on paper, they're there forever, and don't need any realization. But a staggering amount of people were involved in the creation of my Symphony No. 1 (Three Movements for Orchestra), for which I won the Pulitzer. There was the Guggenheim Foundation, which helped sponsor it; the MacDowell Colony, where I wrote the beginning; and, of course, the American Composers Orchestra, all of whom put their collective faith into my symphony and allowed me the time to complete it. Now go back a hundred years and compare the situation: Nothing of the sort could possibly have happened, because society simply didn't recognize female achievements."

Zwilich compares the beginning of a new composition to the armature of a sculptor—a general sort of skeletal mold, with a few details in order. She begins with a substantial amount of musical material, sketched over a period of time—motives, themes, harmonic, structural and dramatic ideas, and a vague conception of form. Then she fleshes the music out, improvising on piano and violin and working with the notes on paper.

Inspiration provides raw ingredients and then, as Zwilich works with what is already written, the music evolves. For all composers, there is an interplay between inspiration and craft.

"Inspiration engenders product, which, in turn, engenders more inspiration," Zwilich says. "All the written arts work this way; you can't imagine an Immanuel Kant in a preliterate age. While still a child, Mozart already possessed an unparalleled musical imagination but his technique evolved over the course of his lifetime, so that his later works have not only the spark and impulse of the early music, but a complexity and depth that can only come from years of experience.

"But it is not enough to manipulate abstract forms and ideas. A composer must also provide color, thrust and purpose, allowing a work to unfold gradually over a given length of time. As such, composition is both a written and a performing art—it must *sound*.

"I become as fully acquainted with a new composition as possible before I begin writing," Zwilich says. "For example, when I began my Double Quartet, I went to Alice Tully Hall to find out exactly where the players would be sitting, and how their sound would reach the audience. By the time I actually started composing, I didn't have to just jot down a disembodied, arbitrary B-flat. I knew that it was a specific B-flat for viola, in this register or that, this context or that. In the case of the Double Quartet, I even knew whether it would sound from the right or the left side of the stage.

"This simply isn't an octet. I wanted the audience to realize that this was not just a piece for two cellos, two violas and four violins, but two separate string quartets, simultaneously competitive and cooperative. Chamber music demands a continual trade-off in musical hierarchy. Now the first violin has the melody, now it's taken by the cello; now the one quartet leads the way, now the other."

Each of Zwilich's compositions has a unique genesis. "I heard the first 15 bars of my symphony immediately," she says. "I never have to work for themes; they can hit me at any time and usually do. When I started the symphony, however, I knew that I wanted to create something that would exploit the rich sonorities of the American Composers Orchestra, and felt free, for example, to write a virtuosic tuba solo, because a modern symphony orchestra is really a stage full of virtuosos. I also wanted to follow an obsession with the minor third—to take this simple interval and use it to generate an entire piece, to create a rich, harmonic palate and a wide variety of melodic gestures, all emanating from a simple source. All the complex harmonies in the symphony are created by piling third upon third upon third."

Still, Zwilich's work is only partly systemic. There are many ways of composing music—from relying on a system so complete that one can explain, technically, exactly why every note is where it is, to a purely intuitive manner of stringing melodies together. Zwilich charts a careful middle course: "I believe in a lot of analysis early in the game, which I then consciously abandon and run wild.

"Once I'm well into a piece, it starts to feed back into me. It's like being a playwright. You create the character, of course, but after a while, the character has a life of his own. By the time Act I is over, you pretty much know what George would or would not say to Martha, and you proceed accordingly, with some sensitivity to dramatic context. After a certain point, a composer's job is to be a good listener—you listen to what's gone before, and then decide how to continue.

"Some things never get easier. The challenge in composition—in all of the arts—is to be a whole person, not merely part of a school or a reaction to the previous generation or century. And, no matter what your track record, when you try to do something you've never done before, you risk falling on your face. So you have to work up courage. But I have this drive—it's sometimes an uncomfortable feeling, almost like an itch, but it's been my best friend, keeping me going.

"There's a funny story behind my string trio. Somebody I hadn't seen in quite a while called me, and asked if I would be interested in doing a work for an upcoming concert. I said that I didn't have the time, but we kept talking for a while and during the conversation I started to hear music. The trio was beginning to take shape in my head. So I said I'd think it over, and I did, throughout the rest of the day and, I guess, through the night as well. In any event, I woke up the next morning, and the whole opening section was waiting to be written down."

The tree, unfettered, begins to dance.

The New York Times Sunday Magazine (1985)

PART II

REVIEWS

1 THE COMPLETE WORKS OF ANTON WEBERN CONDUCTED BY PIERRE BOULEZ

Since his murder in 1945, there has been no shortage of debate on the stature of the composer Anton Webern. At one point, he was considered to be the center of the avant-garde. Musical conservatives hated him, and insisted that his music was nothing more than timid noise. Musicologists admitted to puzzlement at his "gnomic, over-delicate" works. Yet, in the long run, Webern was probably done far more harm by his friends.

For throughout the 1950s and 1960s, a generation of bad composers, armed with Ph.D.s, celestas and crypto-Marxian blather about the "historical inevitability" of 12-tone music, surged through the conservatories blowing the (muted, pianissimo) trumpet of Anton Webern. Tonality was dead, they screamed, forgetting the words of Arnold Schoenberg, the founder of 12-tone music, who cautioned his students not to forget there were many pieces waiting to be written in C major. And music became a protracted game as the academy boys sat in their offices, juggling tone rows, making their calculations and producing some of the dreariest music ever written. All in the name of Anton Webern.

Tonality, of course, was not dead. On the contrary, it is the 12-tone school that has become history. Composition students still write 12-tone pieces the same way they write fugues and passacaglias. But the style is no longer the vanguard of contemporary music. As Jack Beeson recently observed, today's avant-garde is often tomorrow's conservatism, and vice versa. He cites Virgil Thomson's collaboration with Gertrude Stein, *Four Saints in Three Acts,* a work that was considered (musically at least) hopelessly old-fashioned (read "tonal") when it premiered in 1934. Today it seems revolutionary compared to most of the music produced in the 1930s, its absurdist nature fresh and exciting.

Do not misconstrue what I am saying; there is much that is extremely valuable in the 12-tone school. Webern, in particular, impresses me as an undoubtedly great composer, one of the century's finest. He expresses more with a few pianissimo quavers than most composers do with a full orchestra and an hour of time. Yet apparently he cannot be copied; his work is too iconoclastic and personal for easy assimilation. His imitators mimic all of his "tricks" (the sparseness of orchestration, the elegant melodies, the long periods of silence) yet reproduce none of the magic. Webern will be remembered, I think, as an inspired, brilliantly talented eccentric, in the tradition of Berlioz, Pfitzner, Busoni, Alkan and a few others who stand a step away from the mainstream.

Webern's music is highly organized, meticulously arranged. A Webern score can be analyzed so minutely that it could take days to follow the path of his 12-tone rows as they vary and develop. His music is full of inverted canons, row transpositions and other complexities which must fascinate certain kinds of theorists. But these are games, one must assert, and the ultimate test of a work of art is its aesthetic appeal. And there is a tremendous amount of sheer beauty to the music of Webern, much more than this seemingly compulsive ordering. The unique timbres that make up his sound, the haunting power of his minute melodies (now *here* is a minimalist composer!), the economy of his thought—all of these elements are integral parts of the appeal of Webern, and can be appreciated by the layman as well as the musicologist.

Pierre Boulez has finished his 10-year project of recording the complete Webern. It is a magnificent achievement. Besides Boulez, the set features Heather Harper, the John Alldis Choir, the Juilliard Quartet, Charles Rosen (a great classicist who, not surprisingly, reveals himself as a sympathetic and intelligent interpreter of the modern idiom as well), Isaac Stern and the late Gregor Piatigorsky. Halina Lukomska is superb in the Cantatas, making the most of Webern's jagged, dissonant, yet somehow always highly singable vocal lines.

But the great triumph belongs to Pierre Boulez, and his work is a revelation. Boulez is, of course, a well-known composer as well as conductor, and he created some of the better so-called "Post-Webernian" music. He conducts technically "perfect" performances of the scores, of course, but he also imparts a warmth, even a romanticism to the music that most conductors have ignored. One can hear echoes of Mahler, Bruckner and even Johann Strauss (whom Webern adored) in this music.

The Boulez set contains all the music Webern allowed to be published in his lifetime—four short records. (A future set will contain many unpublished works.) As a bonus, we have Webern's orchestration of his own Opus 5, Five Movements for String Quartet, his amazing "Webernization" of Bach's "Ricercata" from the Musical Offering (almost like hearing one great composer through the ears of another) and, finally, a glimpse of Webern the conductor: the great man himself conducting his own arrangement of Six German Dances by Schubert, recorded with the Frankfurter Funkorchester in 1932. This last performance is especially valuable in that it shows how loosely and lyrically Webern conducted an orchestra.

Boulez gives a fine performance of Webern's Symphony, Opus 21. At nine minutes, it obviously bears little resemblance to the epics of Mahler and Bruckner, but is an exquisite, near-ecstatic piece of music—among the most original works Webern ever created—with an emotional expression that belies its length. The Passacaglia, an early, tonal work (and, it must be admitted, a rather bland one), is given a superb performance and contains 15 seconds which, through some odd coincidence, could have come straight out of George Gershwin—a composer who was seven years old when the piece was written. Charles Rosen manages to make a musical statement out of the Variations for Piano, one of the most difficult pieces in the repertoire to pull off correctly. The two late cantatas, with texts by Hildegard Jone—possibly Webern's greatest works—are given first-rate performances that accentuate their mysticism and lyrical bent. But it is purposeless to mention all of the extraordinary moments on this set; everything is at a very high level, and each listener will be able to choose moments that have specific appeal.

There *are* a few minor, extramusical annoyances. The booklet that accompanies the set was obviously cheaply and quickly done. It contains a number of mistakes, notably a discussion of a set of posthumous pieces that was not included in this recording. At the same time, there is no discussion of the string orchestra transcription of Opus 5. There is no mention of the orchestrations of the individual pieces, a serious omission. However, the booklet does contain an informative cross section of quotations from Webern on life and art, and gives the listener insights into his music.

This is a brilliant set, one that anyone interested in contemporary music should have. Webern's genius is a quiet, almost unassuming one, but his music can have a profound and exalting effect on an open mind. Webern's biographer, Hans Moldenhauer, once wrote about the composer's love for mountaineering: "Webern was not a fanatic about attaining the highest point. His passion was not for conquest; he wished only to immerse himself in the wonders of nature, and in the stillness of the heights...." Much the same can be said for the man's unique musical perception.

Soho News (1979)

2 TUNNEL HUM

My editor at the New York Times *felt that this article, written in the spring of 1984, was neither a review nor a news item, and so it was never printed. I continue to find the conflict represented here hilarious and so I publish the story for the first time as an example of the unexpected dramas that any working critic encounters.*

"Come to the performance area prepared to generate energy for positive change," the invitation to Bonnie Barnett's "Tunnel Hum '84" began. "Acting as an individual and taking personal reponsibility for your sound and movement, occupy the space, make yourself at home, gather your energy, build your energy with the goal of pooling your individual energy with the others also working."

And so, on Sunday afternoon, roughly thirty people gathered in the auditorium at P.S. 41 in Greenwich Village, to hum, chant, moan and sing together, in spiritual congress with similar groups in San Francisco, Seattle, Melbourne and London. In fact, the San Francisco, Seattle and New York "Tunnel Hums" were linked by radio; several of the participants at P.S. 41 were tuned in, via portable headsets, to this larger picture.

The atmosphere was initially that of gentle, roseate Utopianism. Everybody hummed around a shared tonal center, changing vowel sounds now and again. Miss Barnett, speaking from San Francisco, affirmed her belief that "singing is vibration, vibration is movement and movement is change.

"In ancient times, the arts—singing, dancing, chanting—were integrated into daily life in a manner that we cannot imagine in our era filled with automobiles and computers," she added, wistfully.

This blissful communalism was suddenly disrupted by one defiant individualist. Richard Vitielo, who later said he was a performance artist with a Master of Fine Arts from Yale University, began to wander around the auditorium waving his arms, practicing what seemed to be tai-chi maneuvers, imitating group members in loud, shrill descant, and generally calling attention to himself. Mr. Vitielo was first asked to calm down and join the group, then to leave, and was finally bodily ejected from the room, howling something about violence, creative expression and civil rights while the vocalists warily hummed accompaniment.

"I was just following my own definition of what they were doing onstage," Mr. Vitielo said later. "It is my gift to be the central person in any performance. I am always the conductor, the leading man, the *danseur*. My art is involved with transformative processes and has an ennobling intent; it is Apollonian."

"This is a group event and cannot be destroyed by the whim of one individual," said Steve Rathe, who was producing the "Tunnel Hum" radio broadcast. "We've worked very hard on this. The whole idea behind Bonnie's piece is to blend and synchronize with other voices around the world. This individual was intent upon spoiling everything, throwing out pitches that were totally irrelevant."

"This really stinks," said a woman who walked out after Mr. Vitielo's eviction. "I don't like those people onstage. They take themselves so seriously. They're full of peace and love but they want to run the world."

"Self-expression is a part of peace," said her companion. "To participate in a gathering like this and then to be told that you're humming the wrong way isn't very peaceful."

(1984)

3 A SUZUKI ORCHESTRA

Onto the stage of Carnegie Hall they stepped, sixty young musicians between the ages of four and sixteen, armed with miniature violins. Friday night's concert by the Haag-Leviton Academy of Performing Arts—under the joint direction of Julian Leviton and Betty Haag—and the Chamber Orchestra, conducted by Henry Mazer, was a celebration of Shinichi Suzuki's controversial pedagogical methods.

Briefly, Mr. Suzuki and his disciples believe that any child can develop a considerable level of musical achievement when taught by rote at an early age. The child does not learn to read music until much later, if at all; the handling of the instrument, playing by ear and unthinking repetition come first.

The students in these orchestras have played throughout Europe, appeared on the television program "Good Morning America" and the French television version of "That's Incredible." Some of the children are only two or three feet tall; all were dressed uniformly, obeyed their conductors, stared straight ahead and bowed with startling precision, offering renditions of everything from Bach and Handel through Bartok and Leroy Anderson. There was something initially arresting about this vision.

However, while nobody expects profundity from preschoolers, the collective playing was automatic. More disturbingly, some solo performances by senior members of the ensemble—pianists and violinists old enough to know better—were technically assured but absolutely devoid of germinal musical insight.

All of which makes one suspect that the Suzuki method is to music making as parrot chatter is to oratory. This listener, for one, would rather hear a fumbled, halting rendition of the Brahms lullaby or "To a Wild Rose" that a child has cherished, wrestled with and made his own than empty precocity on the stage of Carnegie Hall.

The New York Times (1984)

4 "POSTCARD" FROM JUILLIARD

There have been many stagings of Dominick Argento's *Postcard from Morocco* since its first performance in 1971, but it is doubtful that any of these have rivaled the recent Juilliard American Opera Center's production—which this reviewer saw Friday night—for originality, professionalism and sheer visual beauty.

Postcard from Morocco, which Mr. Argento set to a libretto by John Donahue, belongs to a proud American tradition: It is an opera, like *Four Saints in Three Acts* by Virgil Thomson and Gertrude Stein, or *Einstein on the Beach* by Philip Glass and Robert Wilson, that has no clearly discernible plot but makes its effect through a powerful series of images and inferences. It is difficult to say exactly what "happens" in *Postcard from Morocco* but over the course of the opera's hour-and-a-half duration, Mr. Argento addresses several subjects, among them confusion, human cruelty, a clash of cultures and the mixture of pain and perception that go into the making of an artist.

Mr. Argento's idiom is largely tonal, conservative yet distinctly his own, and borrows from ragtime and other strains of popular music. He writes sympathetically for the voice; there are few of the jagged, angular leaps and bounds that so often typify modern opera. He is also a deft parodist—a ballet sequence entitled "Souvenirs de Bayreuth" provides one of the funniest sendups of Wagner since Emmanuel Chabrier turned *Tristan und Isolde* into a *galop* for two pianos.

The action of "Postcard from Morocco" takes place in an outdoor railway station where a disparate group of passengers are, apparently, waiting for a train. Photographs are snapped, a child watches the world with innocent eyes, puppets and marionettes act out the battle of the sexes, a woman sings the praises of her compact mirror, while another delivers an odd torch song entitled "I Keep My Beloved in a Box." Connections are made, and a train snorts into the station (here Calvin Morgan's sets inevitably recall the chugging locomotive in *Einstein on the Beach*). Talking heads pop up from the sand, and toward the end of the opera, the passengers suddenly turn on the gentle Mr. Owen, the man with a paint box, as he is described in the libretto. It makes for a strange, admittedly stylized, but often beautiful evening of theater. The Juilliard production was sensitively conducted by Ronald Braunstein, directed by David Ostwald and choreographed by Francis Patrelle, and featured performances by Jeanine Thames, Korliss Uecker, Vanessa Ayers, Sidwill Hartman, Joseph Wolverton, Peter Gillis, David Stix, Peiwen Chao and Matthew Usman Webber—all uniformly excellent.

Postcard from Morocco has its weaknesses: It is often arch, rather self-consciously clever at times, and may prove, for those who demand linear continuity from theatrical events, somewhat befuddling. But it is rich in ideas, dares to take some chances, and finally, as was once said of modern poetry, adds to our stock of available reality.

The New York Times (1985)

5

A VISIT TO CALIFORNIA

Aptos, Calif.

The drive down the coast from San Francisco is one of the country's most scenic—a panorama of towering conifers and wildflowers, beaches and craggy palisades, all united by the timeless ebb and flow of the Pacific Ocean. It is said that 90 percent of California residents live within ten miles of the water; seen or unseen, the presence of the Pacific is always *felt*, to a degree that is beyond the ken of most Easterners.

For the last two decades, this section of California, with its peerless natural beauty, has also served as a center for human creativity. The Cabrillo Music Festival, which began last Thursday and will continue this week in the village of Aptos, a few miles south of the bustling beach community of Santa Cruz, is now in its twenty-third season. This year, a distinguished group of artists—among them the composers Charles Wuorinen and Elliott Carter, the New World String Quartet, the Masterworks Chorale, the violinist Maryvonne LeDizes-Richard, and the conductor and pianist Dennis Russell Davies, who also serves as Cabrillo's music director—has put together a festival devoted, in the main, to Soviet and American music; the sole exception will be a concluding performance of Beethoven's Ninth Symphony scheduled for Sunday.

No festival, no matter how well intentioned or carefully planned out, could begin to sum up the vast diversity of American music in sixteen concerts, and we are still far from an understanding of what is really going on, musically, in the Soviet Union. There is nothing comprehensive about this festival, nor could there be. That said, the Cabrillo offering allows the listener a tantalizing sampler. Although, in the spirit of a festival, shared attributes rather than differences between the United States and the Soviet Union were stressed, there was, happily, no attempt at a false equation of the two countries: It was pointed out, for instance, that several Soviet composers were invited to participate but were denied permission by their government, and that certain members of the Moscow and Leningrad avant-garde had been officially censured over the years for what was perceived as their experimentalism.

The festival began with "Four Ragtime Dances" by Charles Ives—

an unfortunate choice. Ives has always impressed this listener as a week-end dabbler who, very occasionally, hit upon a good idea; this mash of jazz rhythms, polytonality and "Bringing in the Sheaves" did not seem to be putting our best foot forward.

Most of the homegrown pieces included in the festival are more rewarding—works by Leon Kirchner, Lou Harrison, David Diamond, William Schuman, Hall Overton, Aaron Copland, and the early American-Moravian composers John Antes and John Frederick Peter. Mr. Carter and Mr. Wuorinen, the festival's composers in residence, are represented by several large works—Mr. Carter by his String Quartet No. 2, the Double Concerto, the Variations for Orchestra and the West Coast premiere of his Triple Duo; Mr. Wuorinen by the West Coast premiere of his "Winds," and "Bamboula Squared" and "The Magic Art: An Instrumental Masque Drawn from Works by Henry Purcell."

The latter work was especially impressive. Mr. Wuorinen has not attempted a "Pulcinella" here—there is no sense of a contemporary composer dolling up a past master in the best Modernist finery. Indeed, Mr. Wuorinen approaches Purcell's chaste harmonies with genuine reverence, and saves most of his "touches" for orchestration—the piccolos emulating a steam whistle here and there, the muted trumpet snarling a fanfare. The finale was delightful: a busy, colorful rave-up that called Mr. Copland's work to mind—"Dido and Aeneas Go to the Rodeo."

Mr. Wuorinen's intelligence is never in doubt, but one admires works like the "Percussion Duo" (1979), which was performed on opening night, from a certain distance. There is little sense of warmth or sensual beauty; it recalls a brilliantly fashioned ice sculpture. Whatever rewards it provides can be measured as a quickening of the mind rather than of the pulse. One wishes Mr. Wuorinen had listened closely to 20th-century masters like Richard Strauss and Jean Sibelius as well as his cherished Schoenberg and Stravinsky. The performance, by the percussionist William Winant and the pianist Emily Wong, was clean and rhythmically charged.

Other worthy performances in the first days of the festival included Charles Dowd's virtuosic rendition of Mr. Carter's "Four Pieces for Four Timpani," Miss LeDizes-Richard's impassioned performance of Mr. Schuman's Concerto for Violin and Orchestra. The Soviet pieces heard in the festival's first concerts were not particularly prepossessing. Aleksandr Knaifel's "Da," although described as mini-

malist, had more in common with the sparse structuralism of Morton
Feldman than the hyperkinetic repetitions of Philip Glass. Sofiya Gu-
baidulina's "Concerto for Bassoon and Low Strings" was an evocative
combination of extended woodwind techniques, and a more conventional lyricism; it was sensitively played by Gregory Barber. Other
upcoming concerts promise a representation of the great late chamber
works of Dmitri Shostakovich—written after he had finally escaped
state-ordained Nationalist tub-thumping—and compositions by Sergei
Slonimsky, by the arrestingly eclectic Alfred Schnittke and the more
conservative Rodion Shchedrin.

Full houses greeted the concerts warmly. There seems to be a great
deal of support for the Cabrillo Festival in and around Santa Cruz, and
it is well deserved.

The New York Times (1985)

6 JOHN CAGE AT BARNARD

This listener once attended an amusement park sideshow devoted to
the life and adventures of Elsie the Cow. The gentle beast, by then quite
advanced in age, sat placidly upon the stage, eyeing the audience with
serene detachment. An announcer's voice described Elsie's past travels
throughout the world while papier-mâché images of London, Paris and
the Far East were assembled and dismantled around her. Except for an
occasional twitch of the tail, Elsie never moved.

In no way am I equating the distinguished American composer and
philosopher John Cage with Borden's distinguished American cow, but
Opera Uptown's imaginative presentation of "The Bus to Stockport and
Other Stories" at the Minor Latham Playhouse on Saturday afternoon
had the same sense of a stationary central character around which the
world revolved. The production might have been subtitled "Variations
on the Theme of John Cage"—it was a 40-minute theater work constructed around the sounds, ideas, stories and even the physical presence
of Mr. Cage.

"The Bus to Stockport," which featured music by Eric Valinsky and Peter Schubert, a text by Mr. Cage and direction by Rhonda Rubinson, had no real plot and offered no tidy aphorism to take with us when we left. Instead, it seemed a felicitous working-out of Mr. Cage's ideas about random order and indeterminacy.

Mr. Cage, clad informally, sat on one side of the stage and read some charming, anecdotal pages from two of his books, *Silence* and *A Year from Monday,* in a soft, dispassionate voice. A good deal of seemingly unrelated activities revolved around Mr. Cage; he was not amplified and he often allowed his narrative to become lost. Singers came and went, singing snatches of madrigals; a bass player sawed some deep Modernist continuo; percussion instruments chimed and rattled; lithe dancers stretched their muscles.

The sets were a succession of frames, platforms and sashes that rose and fell, fell and rose. One quickly abandoned any effort to make linear sense of the play, and followed it instead as a series of isolated moments, alternately dull and engrossing. All of which would, most likely, have pleased Mr. Cage, who once observed that what is boring in one minute may be very interesting in ten.

Indeed, this was a vital, appropriate homage to an American original, one that had been planned with care and sophistication. The score (with what was billed as "additional music by Richard Einhorn and J. S. Bach") did not, to the best of my knowledge, use any of Mr. Cage's own compositions, but relied heavily on his most celebrated invention—the prepared piano. This is a standard piano that has been doctored by placing screws, erasers and other material between its strings. On paper this seems like a cute Dadaist trick; in practice, the piano becomes a percussion orchestra, with an ornate sound that calls the Balinese gamelan to mind.

Although this was presented by the Barnard College Theater Program, there was very little to suggest a student production (I can think of several celebrated performance artists who could take lessons from the triumvirate of Valinsky, Schubert and Rubinson). Credit must also be given to the choreographer Pat Cremins and to Chisa Hidaka and June Omura, her stonily beautiful dancers. The other musicians, all capable, were Michelle Bobko, Donald Andrew Howell, Karen Krueger, Joel Mitchell, Risa Evans, Ann Sera, Michael Taddei, Don Yallech and Robert Stanley.

The New York Times (1986)

BACH AND THE RODENTS

There has been much debate on the correct way to perform Bach cantatas. One approach (which, until Sunday afternoon, seems to have been unaccountably neglected) is to dress actors up in blankets, pig masks and wooden rodent heads, blast air-raid sirens, stage a tentative waltz (in four-four time) between an animated mannequin and a woman made faceless by tangled hair, and build to a grand finale featuring the image of a Sun God that closely resembles a fried egg, and placards of happy third-world peasants lifted from a Sandinista mural.

Well, now it's been done, and it is doubtful that one will ever attend a more successful Bach cantata in the "rodent-head" tradition than the one staged Sunday at the Church of St. Ann and the Holy Trinity in Brooklyn Heights. The work chosen was "Christ Lag in Todesbanden," a cantata for soprano, mezzo-soprano, bass and chamber orchestra that has long been one of Bach's most familiar. From a purely musical perspective, this was a chaste, loving performance, sung with grace and feeling by Dawn Upshaw, Rebecca Russell and Wilbur Pauley, with expert accompaniment by the Arts at St. Ann's Chamber Orchestra and the St. Ann's School Chorus, under the direction of Fred Sherry.

But there were these *rats*, you see, and men in suits with shiny briefcases, and storm troopers with sticks, and smeared, anguished faces that could have been borrowed from the triptychs of Francis Bacon— all courtesy of the Bread and Puppet Theater, under the direction of Peter Schumann. The music itself was not disturbed (during arias and choruses the troupe limited itself to wordless dramatics) but between movements all was fair. So one heard shrieking and sirens, what sounded like an accordion being played out of tune, and the hard disturbing sound of wooden staff meeting platform.

The Bread and Puppet Theater, formed in New York and currently based in Vermont, attempted no literal representation of the text but instead built a collection of disturbing images. There was much that seemed silly and simplistic (how did the pagan Sun God invade such a resolutely Christian cantata?) but the troupe did convey a certain fearful agony of spirit that was not inappropriate to the work.

One hesitates to discourage creative rethinking of the classics, but Bach's great music, unadorned, speaks with far greater urgency than Mr. Schumann's nightmare skit could begin to convey. This seemed, for the most part, a rehash of the street theater of an earlier generation.

The remainder of the program was less ambitious but more rewarding. Stravinsky's "In Memoriam Dylan Thomas," an austere setting of the poet's "Do Not Go Gentle into That Good Night," began the afternoon, sung, with grave dignity by Mark Bleeke. Mr. Sherry was the cello soloist in Tchaikovsky's own orchestral adaptation of the "Andante Cantabile" from his String Quartet (Op. 11). He played with patrician elegance and without undue sentiment. Dapper readings of brief Renaissance compositions by Gibbons, Coperario and Bull rounded out the program.

The New York Times (1986)

8 THE RETURN OF CONLON NANCARROW

New York welcomed the composer Conlon Nancarrow back to America after a forty-year absence with a packed house and a standing ovation at Alice Tully Hall.

Mr. Nancarrow has spent the majority of his career fashioning works for the player piano, but anyone who attended Saturday night's presentation expecting dulcet parlor melodies along the lines of "Tiptoe Through the Tulips" or "Tea for Two" would have been disappointed. Mr. Nancarrow, who meticulously punches every piano roll himself, often working as long as nine months on a five-minute composition, has created an invigorating needle shower of sound, replete with jerky, angular rhythms, densely contrapuntal and of startling complexity, played at a speed that would be beyond any human pianist (or any five human pianists, for that matter). He is, in fact, one of our pioneer electronic composers; in his hands the player piano has become a lyre for the brave new world.

As it happened, the concert, produced by the new-music ensemble

Continuum, with guest appearances from the violinist Marilyn Dubow, the pianist Yvar Mikhashoff, a small chamber orchestra and the Cassatt String Quartet, concentrated almost entirely on Mr. Nancarrow's work for live musicians. The player piano was conspicuous in its absence (although two studies were heard on tape, and most of the other works played were transcriptions).

The evening was a rewarding one. Mr. Nancarrow, who has lived in Mexico since experiencing what he considered political harassment from the United States Government in 1940 (he had served in the Spanish Civil War International Brigade), is an American original. He has mapped out a corner of the musical world that is his alone; none of the works heard—from the early "Prelude and Blues" (1935) to "Piece No. 2 for Small Orchestra" (1986)—could have been from any other composer's pen.

Part of the reason for Mr. Nancarrow's success is his frank realization of his limits. He does not write memorable melodies, and his ear for timbre is not impressive. But his music has a wild, spiky energy, an irresistible rhythmic propulsion and enough formal rigor to delight any good Modernist. There is a craggy strength to his music; as with the works of such Americans as Carl Ruggles and Elliott Carter, you may like Mr. Nancarrow's music or you may not, but it is both personal and specific.

As usual at Continuum concerts, the performances were beyond praise—spirited, scholarly and committed. Mr. Mikhashoff's pointillist transcriptions of player-piano studies for chamber ensemble were apt; indeed, this listener found them more interesting than Mr. Nancarrow's new composition for orchestra. A high point of the evening was the attempt by Joel Sachs, who shares Continuum directorial duties with Cheryl Seltzer, to engage the taciturn Mr. Nancarrow in a mid-concert interview. Although the composer tended to answer in one or two words, the interview was received warmly by the audience.

The New York Times (1986)

The composer Milton Babbitt turned seventy last week, and there has been no dearth of commemoration. Still, it is unlikely that Mr. Babbitt will receive a more earnest, loving, intelligent and appropriate tribute than the pianist Robert Taub's recital Wednesday night at Alice Tully Hall.

Mr. Taub, who has recorded all of the composer's piano works for Harmonia Mundi, did not limit his program to Mr. Babbitt, nor did he confine his work to a little "avant-garde ghetto" at the beginning or end of the evening. Instead, he began with Bach, moved on to Babbitt and Brahms, then (after an intermission) to more Babbitt, and he concluded with Ravel's "Gaspard de la Nuit." A musical continuum in no chronological order, but aesthetically satisfying.

Mr. Babbitt's music is too often treated like some rare form of hieroglyph—unquestionably important but indecipherable. Mr. Taub's written introduction to his program, which flies in the face of all received wisdom about the composer, deserves quotation: "The music of Milton Babbitt must be played from the heart. The dazzling, highly imaginative pianism—enormous registral leaps, juxtaposition of dynamic extremes, highly complicated rhythms, innovative pedal techniques—always serves an intensely musical end which, as in all great works, should be so completely mastered that the music is free to soar in performance."

Mr. Babbitt's fearsome reputation has been earned more through his words than his music. He still receives an occasional angry letter about his 1950s essay in *High Fidelity* magazine, which he called "The Composer as Specialist" but which the editors endowed with the more provocative—albeit misleading—title, "Who Cares If You Listen?" The phrase was calculated to inflame, and it strengthened a public image of the modern composer as a chilly, clinical manipulator without song or soul.

It must be admitted that Mr. Babbitt has not always helped his case. He chooses titles like "Partitions," "Post-Partitions" and "Canonical Form" and, in one program note, he refers to "models of similar, interval-preserving, registrally uninterpreted pitch-class and metrically-durationally uninterpreted time-point aggregate arrays." Yes, this all means something, and something important, but it is a specialized language addressed to other composers with preoccupations like those of Mr. Babbitt, and need not concern the lay listener.

What I find impressive about Mr. Babbitt's music is its ethereal spring, its lambent, shimmering grace, which he seems to have distilled to its purest essence. The "Semi-Simple Variations" (1957) last barely a minute but present five distinct variations on a compressed theme. In "Lagniappe" (1985), given its concert premiere, one finds the hyperactive glitter of the early compositions combined with a new space and dramatic forcefulness. All of his work is fiendishly difficult, notes splashing out in every direction, seven or eight dynamic levels within the course of one rapid measure.

Mr. Taub's pianism on Wednesday was exemplary. The "Quattro Duetti" by Bach that opened the program were played very simply, the two in intertwining voices essayed with a chaste intensity. Three late piano works by Brahms made a poetic impression: They seemed little daguerreotypes in sound, period pieces in the best sense of the word. And "Gaspard" was played with a Classicist's sense of structure. Other pianists have brought more terror to the central movement, "Le Gibet," with its menacing stasis, but few have unified the work with such consistency. Mr. Taub plays Mr. Babbitt's work with more than professional affinity, he plays with love.

The New York Times (1986)

10   *"BLEECKER STREET" IN CHARLESTON*

Charleston, S.C.

This gentle, antiquated coastal city is throwing its annual party. On Friday, the Spoleto Festival USA opened its tenth season here, and talk of opera, dance, mime and theater has supplanted the traditional topics of gardens, squalls, Civil War and she-crab soup. This year, the festivities have an additional impetus: 1986 is the year that Gian Carlo Menotti—the composer, director, librettist, Spoleto founder and resident guru—turns seventy-five.

It seemed fitting that a revival of Mr. Menotti's opera *The Saint of Bleecker Street*, which won the Pulitzer Prize for composition in 1954, should open the festival. It is not the composer's most popular opera—*Amahl and the Night Visitors* must take pride of place. Nor is it his most visceral: *The Consul* and *The Medium* are both more gripping, more compulsively engrossing theater works. But "The Saint" is probably the most esteemed.

The Saint of Bleecker Street is set in New York's Little Italy, and contrasts the mystical visions of a dying girl, Annina, with the cynicism of her brother, Michele, and the worldly love of his mistress, Desideria. Eventually, through malice and misunderstanding, all three are destroyed, although Annina is beatified as she dies. A lot of subjects are skimmed over—intolerance, Italian nationalism, the line of demarcation between sickness and vision, and the particular pain of being gifted and therefore different. Mr. Menotti's libretto may not offer much in the way of philosophy, but he knows how to tell a story, and the story is often moving. His music is tonal, frankly emotional, serious—and uneven. Its best moments—the rattle of an orchestral subway, the terse, cogent interludes, virtually all of Act I and the final apotheosis—elevate *The Saint of Bleecker Street* to a place above Mr. Menotti's more popular operas.

The Spoleto staging (a co-production with the Opera Company of Philadelphia) has been handsomely mounted. Zack Brown's sets—depicting Bleecker Street flats, vacant lots in Little Italy, an Italian restaurant, and even a subway station—have the gritty detail of Weegee still lifes. Christian Badea, who conducted the youthful Spoleto Festival Orchestra and Westminster Choir, led the opera as one unbroken arc of sound. He seems to genuinely love the score, and his forces played with accuracy and sweep. Mr. Menotti himself was responsible for the stage direction; we may presume that it is definitive.

Gail Dobish made a splendid, fully dimensional Annina: her voice now strong and ringing, now delicate and filled with childlike wonder. Franco Farina's Michele was also a vivid portrayal. Most of his role is declaimed in a Menottian *Sprechstimme*, but he sang his occasional aria with palpable ardor and a versatile baritone voice. Leslie Richards made a husky, arrestingly physical Desideria, while Julien Robbins brought a paternal dignity to the role of Don Marco. There is effective support from Anna Maria Silvestri, Margaret Haggart, Antonia Elisabeth Brown and David Barrell.

One hopes Mr. Menotti will someday write an opera that has no screams, no stabbings, no crimes of passion, no flouncing effusive arias and no dynamic levels above a temperate mezzo-forte. Mr. Menotti is, by all accounts, a witty, cosmopolitan man. One wishes his operas reflected this sophistication more directly. He could use a sort of musical Maxwell Perkins to calm and focus his aesthetic, to point out, say, that Act II is tawdry and obvious in comparison with the acts that surround it.

Yet despite its vulgarities, *The Saint of Bleecker Street* remains a convincing opera, written in a manner that is organic, individual, accessible and idiomatic for the voice. It is now more than thirty years old and shows little sign of losing its power. In fact, it is probably more successful as a period piece than it ever was as a contemporaneous "slice of life." One suspects it will endure.

The New York Times (1986)

11 PAVAROTTI AT THE GARDEN

Luciano Pavarotti swaggered onto the stage of Madison Square Garden Tuesday night like a conquering heavyweight, arms lifted high above his head, a 360-degree grin on his face. Television cameras whirred and diamonds flashed in the huge auditorium, which was more than half full. This was, by any standards, a major event—an event, however, that had very little to do with music.

It is important to be fair about this. One need not have been bewitched by the deluge of publicity Mr. Pavarotti has received in order to admire his singing. Whether or not he is "the world's greatest tenor," as common wisdom would have it, seems a small matter, but he *does* possess a voice of surpassing sweetness, artistic gifts of no little dimension and a vivid and likable stage persona. The puffery and hype that surround Mr. Pavarotti are unpleasant; the career choices he has made are regrettable. But there are some solid musical reasons for the adulation.

Mr. Pavarotti chose a standard selection of tenor arias—from

L'Elisir d'Amore, Mefistofele, Werther, Tosca, La Bohème and *La Gioconda*—
and added "Dal piu remoto esilio" from Verdi's *Due Foscari* by way of
novelty. The singing was, for the most part, slick, eminently professional
and dispassionate: Mr. Pavarotti must have sung "Che gelida manina"
several thousand times, and the routine is starting to show. Most of the
selections were tossed off with a Vegas breeziness: "Giunto sul passo"
from *Mefistofele* was no longer an aged philosopher's satiated lament,
but merely a celebration of a healthy tenor voice. He was at his best in
the two selections from *Tosca*—"Recondita armonia" and "E lucevan
le stelle"—which were sung with a compelling urgency.

Still, one questions why anyone would want to pay the sort of prices
that Mr. Pavarotti now commands—the top ticket last night was $75—
to listen to him sing in Madison Square Garden. There is undoubtedly
a certain excitement to being in the physical presence of an artist one
admires. But Madison Square Garden has more than 15,000 seats, so
intimacy was impossible. In addition, Mr. Pavarotti was, by necessity,
amplified, and his voice lost its natural lustre and took on a hollow,
machine-like quality. For $75, one can purchase many hours of Mr.
Pavarotti's singing on LP or compact disc, captured faithfully for repeated hearing in the comfort of home.

Dame Joan Sutherland had been scheduled to sing with Mr. Pavarotti, but bowed out a little over a week ago because of an ear infection.
The soprano Madelyn Renee, who sang in her stead, has a charming
stage manner and a voice of darkish hue and moderate size and strength,
but she is not ready for this sort of exposure, and such premature display
does her no good service. Can Miss Renee really have been the most
qualified soprano willing and available to sing with Mr. Pavarotti on a
week's notice? The mind boggles.

The New York Times (1986)

12 PIANO AND ELECTRONICS: REBECCA LA BRECQUE

Rebecca La Brecque's Sunday evening recital at Merkin Concert Hall was one of the season's most challenging. While it did not add up to a fully successful program, it may have been a historic one, for it contained no fewer than seven world premieres written for a combination of media once thought to be mutually exclusive.

Ms. La Brecque is best known for her pianism: She has recorded the three piano sonatas of Roger Sessions and several compositions by Frank Martin. For Sunday's recital, she commissioned seven new works for piano and Yamaha DX7 Synthesizer—with both instruments played simultaneously—by William Kraft, David Gottlieb, Gregory Reeve, David Froom, Dean Friedman, Conrad Pope and Robert Pace.

When the synthesizer first began to win general popularity, vast claims were made for its range and versatility. Many of these claims were couched in the negative—I recall articles that, with apocalyptic glee, prophesied the end of the symphony orchestra, traditional instruments and, indeed, all acoustic music.

With the development of contrapuntal technology and the ability to play more than one line of music at a time, the claims redoubled. And a great deal of progress has been made in developing new techniques for the instrument—for it was, after all, just one more instrument, albeit one of extraordinary promise. We now have all the syntax we need; what has not yet been discovered is a way to use this syntax lyrically, with warmth and humanity.

The masterpieces of electronic music—the "Poème Electronique" by Edgard Varèse, "Song of the Youths" by Karlheinz Stockhausen, "I Am Sitting in a Room" by Alvin Lucier, "Ensembles for Synthesizer" by Milton Babbitt, even the early tape pieces of Vladimir Ussachevsky and Otto Luening—are evocations of strangeness. We are chilled by the Varèse and Stockhausen works, tantalized by the shimmering grace of the Lucier and Babbitt pieces and charmed by the space-age evocations of the Luening and Ussachevsky works. But, in every case, the sense is that of a brave new world rather than the one we live in. Some would say that expression lies outside the range of a "machine" such as the synthesizer. But the piano is a machine, as is the organ; the challenge is to bend the machine to human wishes.

And the challenge was not met, at least by the music played Sunday. Mr. Pope's "Jeux-Partis" was the evening's most immediately gratifying work, full of ideas and imbued with icy grandeur. Mr. Kraft's "Requiescat" began with drones and gentle chiming that gave way to an angry succession of tone clusters. Mr. Gottlieb's "(Re:) Location and Relocation" was a scurrying, concentrated etude. Mr. Reeve's "Orchives" featured the most democratic division of duties between the two instruments: Most of the other pieces seemed either works for synthesizer with piano addenda or vice versa, but here the blend was nearly seamless.

Mr. Froom's Second Ballade combined glassy, brilliant writing for the piano with cloudy clusters from the synthesizer. Mr. Friedman's "Daphnia," informed by a welcome sense of humor, began as an exaltation of cicada chirps, transforming the sound eventually to a sinister whir of helicopter blades. "Night Fantasy," by Mr. Pace, was something of an antiphonal—the piano played, then the synthesizer, then both in a linear progression that eventually led to a fierce, bacchanale-like finale.

As for Ms. La Brecque, she was simply magnificent—a study in concentration, coordination, fury and musicianship, dashing around the stage in a hyperkinetic blur, crossing hands from instrument to instrument, flicking switches, executing the most complicated, mercurial filigree or crashing her arms down on half of the keyboard in a sonic explosion.

The New York Times (1986)

13 SWEENEY TODD AT CITY OPERA

As I wandered through the lobby of the New York State Theater during intermission Wednesday night, I heard a disgruntled matron complaining about the inclusion of a "musical" such as *Sweeney Todd* in the repertory of a "serious" opera house.

A complaint worth addressing, since it has found an echo in high places. Indeed, several critics and arts panelists have recently spoken out in favor of a clear separation between opera—a "high" art—and musical theater—dismissed as "entertainment." The argument has been grossly oversimplified; one critic recently bewailed, in his best Spenglerian manner, the crude sensibility that cannot distinguish between *Fidelio* and *Oklahoma!*

Well, of course. But the comparison is not germane. Our opera houses hardly present a steady diet of *Fidelio;* you are much more likely to run across *I Pagliacci, La Gioconda, Tosca* and *La Bohème.* And those who believe that the operas of Mascagni, Cilea, Leoncavallo, Ponchielli and Puccini occupy some higher moral and artistic plane than the musicals of Cole Porter, George Gershwin, Jerome Kern, Noël Coward, Leonard Bernstein and Stephen Sondheim are, simply, kidding themselves.

Sweeney Todd, for all of its humor and chills—and it is both genuinely funny and genuinely terrifying—is a deeply serious work. Stephen Sondheim's *grand guignol* horror show, about a disaffected barber who dispatches his clients and sells them for meat, addresses several important themes—power, injustice, despair, cynicism, and, however obliquely, love—in a manner that engages and disturbs, entertains and alienates. *Sweeney Todd* has ideas and it has force; the music, barbed, tangy and sentimental by turn, has staying power. To deny it a place in the repertory seems to me nothing more than cultural snobbery. May we begin to appreciate our own, "homegrown" music or must we genuflect to Europe forever?

The City Opera has mounted a production of *Sweeney Todd* that closely recalls Broadway's—all pulleys, ladders and revolving sets. Timothy Nolen made a rather low-key but convincing Sweeney Wednesday: If the character seemed less demonic than it did on Broadway, it also seemed more poignant and multidimensional. Joyce Castle sang Mrs. Lovett in a rich, dark voice; she, too, brought a surprising tenderness to her character, which remained an embodiment of evil, but no longer *cartoon* evil.

Cris Groenendaal, the only carry-over from the Broadway production, did his best with the rather simpering character of Anthony, while Leigh Munro's Johanna was appropriately skittish and highly strung. Will Roy and John Lankston brought the necessary malevolence to the roles of Judge Turpin and the Beadle (the City Opera has restored Judge Turpin's ugly and impedimental self-flagellation scene, which was wisely cut from the Broadway production).

Robert Johanson, in his company debut, managed to find a degree of subtlety in the character of Tobias—a role usually shouted, in the shrillest of tones, handily marring any sympathy the character is supposed to inspire. The beggar woman has two distinct moods, lusty and piteous, and Ivy Austin illuminated both. Jerold Siena was an appropriately hammy but indistinct Pirelli.

I strongly recommend *Sweeney Todd*. It represents the American musical at its most daring and intelligent.

Newsday (1987)

14 MEREDITH MONK AT BAM

"The voice itself is a language," Meredith Monk has written. "It can uncover shades of feeling that we have no words for."

Something similar might be said of Monk's music. I have never fully understood why the best of it moves me so deeply. Her work is simple, ritual, chirpy and chantlike by turn, almost runic in its strangeness and intensity. Stark chords are repeated again and again, singers howl and cavort, shouting syllables that mean nothing and everything.

It is almost context-free, with as much in common with African pygmy music as with any traditional Western modernism. There is jazz in there somewhere, and rock, and a personal minimalism. And Monk's music is informed by her work with dance, by her films, by her experiments with vocal techniques. It all adds up to something unique and original, which demands respect and a certain reorientation of focus.

Monk's current show, which I saw Sunday afternoon at the Brooklyn Academy of Music, is a mixed bag. The new piece, entitled "The Ringing Place," strikes me as dreary and overlong (its placement, at the end of an ambitious afternoon, when quite a bit had already been said and done, may have had something to do with this reaction). Like many of Monk's works, "The Ringing Place" is serpentine in form. One section follows another, and then another, and another, until it stops.

"The Ringing Place" did provide one insight: If serial composers may be said to construct their works in the careful, considered and unrelentingly aggressive manner of a chess game, Monk's impetus would seem to be that of a recess playground. Girls and boys paired off in corners, giggled, squealed and entered into private games without goals or a formal beginning or end but rather a gradual, enveloped and appreciative shaping and unshaping of time and space.

Monk's duets with the jazz vocalist and arranger Bobby McFerrin went on a little too long, but were for the most part delightful; the two babbled and parried *a capella* with humor and charm.

I would be curious to hear what sort of music Monk would produce in a work specifically for children; there were several in the hall on Sunday and they responded to her music with delight. (This is not a putdown, by the way, nor does it mean that Monk's work is not fit for what we used to call "mature audiences." In fact, writing well and without condescension for a youthful audience should be one of the highest aspirations of a composer.) I also suspect that the shorter pieces on the program might find an unusual but satisfying home in a cabaret setting.

As usual, Monk's ensemble performed its duties with diligence, realizing the music with the idiosyncratic combination of humor and high seriousness it deserves.

Newsday (1987)

15 HARRY PARTCH IN PHILADELPHIA

The composer Elliott Carter once said that the music of Charles Ives was more often interesting than good. A similar observation might be made about the work of the late American composer Harry Partch, whose ambitious, ninety-minute music theater work, "Revelation in the Courthouse Park"—a stylized setting of "The Bacchae" by Euripides—received its professional world premiere in Philadelphia on Thursday night.

Partch was an extraordinary character. A truculent, hard-drinking independent who shunned the musical mainstream and lived much of his life in the California desert, Partch fashioned instruments out of surplus airplane fuel tanks, Pyrex chemical jars, artillery shell casings, bottles and old keyboards. He invented his own tuning systems (dividing the scale into quarter-tones and even smaller fractions) and took at least as much inspiration from the percussive, ritualized music of the Far East as he did from Western Europe.

For most of his life, Partch labored in obscurity. Happily, in the last five years of his life, he was "discovered"—not only by his fellow composers but by the general public (CBS recordings of "The Delusion of the Fury" and some smaller works were classical best-sellers for a while). Since his death in 1974, however, Partch's music has received few performances—mainly because there is only one set of instruments in the world on which it can be correctly played.

There may be another reason, for I continue to find Partch's music more satisfying in the abstract than I do in performance. Arnold Schoenberg once told John Cage that he was not a composer, but rather an inventor of genius; in a similar spirit, I am often more impressed by Partch's physical *inventions* than I am by his musical invention.

I also miss a sense of linear discourse in Partch's music—a unified musical language. For example, while I admire the writings of both Samuel Johnson and Tom Wolfe, I would be nonplussed if Wolfe suddenly affected the high Anglicism of Johnson's prose, and downright shocked if I ran across a WHEEEE!!! in the middle of one of Johnson's "Rambler" essays.

Put simply, "Revelation," composed in 1959, is a jumble of ideas, inspired and not, replete with marching band, tumblers, jugglers, a drum majorette and Partch's unusual instruments. The fatal quarrel between Dionysus and Pentheus finds a modern echo in the persons of Dion, a rock idol, and Sonny, a conventional young man; by extension, the dichotomy is between order and orgy anytime, anywhere. The score is noisy, assertive, sparkling and poetic by turn; one of the big ensembles called to mind the rumble scene in *West Side Story*, reduced to chorus and rhythm track.

But the play's tragic essentialism is lost in flash and detail. Carl Orff's settings of Greek tragedy, created at roughly the same time as "Revelation," kept coming to mind. For all of their occasional crudities, these works have the flesh, blood, muscle and bone that one finds always in Euripides and, alas, rarely in Partch.

Still, we owe the American Music Theater Festival our gratitude for presenting this unusual work, and I cannot imagine a much more eloquent or "authentic" performance (Danlee Mitchell, one of Partch's associates, the leading authority on his work and the curator of his instrument collection, was the musical director). Obba Babatunde, as Dionysus and Dion, called to mind a sort of streamlined, sexy Chuck Berry, while Christopher Dunham was a dignified Pentheus. Suzanne Costallos brought a degree of tragic intensity to the dual roles of Mom and Agave, and there was strong support from Edward Earle, Matthew Kimbrough, Casper Roos, Olivia Williams, Rozwill Young and a host of graceful, muscular tumblers, who bounced across the stage as if it were an acre of trampoline.

Newsday (1987)

16 "SWING" AND SWADOS

If a stroll through certain neighborhoods in Berkeley or Cambridge can bring back everything one hated about the 1960s, Elizabeth Swados' *Swing*, now playing at the Brooklyn Academy of Music, summons to mind the most fatuous, slippery and sentimental pieties of the 1970s.

Swados describes *Swing* as a "concert with theater." She has gathered together some twenty-five young musicians and dancers, aged nine through eighteen, shepherded them through a series of workshops, and fashioned a seventy-minute piece of musical theater, along the lines of her 70s hit, *Runaways*.

The process was not, apparently, a smooth one. "The kids of today, as opposed to those I worked with ten years ago in *Runaways*, are quite different," Swados observed recently. "They are more introverted, disillusioned and apathetic. I've had to adjust my technique to a different mentality and sensibility in the youth growing up in our city."

Whatever adjustment Swados made didn't take. Simply put, *Swing* is cant through and through, and I didn't believe a moment of it. The music is second hand, the philosophy pernicious. The fact that *Swing* addresses a serious matter—the plight of children growing up on the streets of New York—only makes it the more offensive.

Swados seems to have learned everything she knows about young people at Greenwich Village cocktail parties, and her dialogue could have been taken from an old Jules Feiffer cartoon. The children raise clenched fists—¡Viva la Revolucion!—and brandish machine guns. Adolescent hostility toward authority is represented as righteous political rebellion rather than the temporary—and unhappy—glandular dysfunction that it is. (Swados would do well to peruse Wyndham Lewis' devastation of "the cult of the child", before she embarks upon her next project.)

The biggest problems facing Our Youth would seem to be "persecution, paranoia" (chanted *en masse*), the existence of expensive hotels, rich people, parents, angry policemen, and that bugaboo of choice, alienation. To spout this da-da-goo-goo sociology in a world beset by fifteen-year-old mothers, delinquent fathers, the threat of AIDS, the decay of city schools, the random violence, the lack of adequate protection and the infestation of drugs is frivolous at best. Swados even tacks on a glib happy ending, as everybody sings "What you want / What you wish / It shall be / It shall be." No evidence is offered to back up this assertion, and the children won't have BAM and Swados to stand behind them forever.

Other absurdities. There is a syrupy recurring theme which could have come right out of a Peanuts TV special (Charlie Brown shuffles through the autumn leaves, hangdog expression on his face). And what is a Bennington-educated woman doing writing a song with a refrain like "God, but life *be* one long pain"? Why does all the jazz sound like mid-70s collegiate MOR? How many ghetto children bop around singing "Spiritus Sanctus, Kyrie Eleison"?

Still, the kids *are* good; against all odds, they performed *Swing* with energy, agility, irrepressible high spirits and an utter lack of pretension. I liked especially the delightfully squeaky girl who wriggled through "12 Rough Years" like a combination of Little Eva and Fanny Brice. To adapt a 70s maxim, it may be that there are no bad kids, only bad playwrights.

Newsday (1987)

Irving Berlin, America's most prolific and versatile songwriter, has reached the age of 100, and his songs show no sign of losing their lustre.

On Wednesday night, an appropriately glittering audience paid up to $1,000 to attend a benefit concert at Carnegie Hall, in honor of the Berlin centennial. The songwriter wasn't there—he celebrated his birthday in the East Side town house he has occupied for many years—but Frank Sinatra was, as were Shirley MacLaine, Marilyn Horne, Morton Gould, Rosemary Clooney, Garrison Keillor, Tony Bennett and many other luminaries from all areas of the entertainment business. The proceeds of the concert, which were expected to exceed $500,000, will benefit the American Society of Composers, Authors and Publishers (ASCAP), of which Berlin is a charter member, and the Carnegie Hall Foundation.

To their credit, the producers made the effort to present Berlin in all of his guises. There were songs from 1911 and from 1962. The cast was multiracial (few, if any, composers, have done so much to bring black music to a wider audience), and we heard representative snippets of Berlin the jazzman, the sentimentalist, the patriot, the teaser, the lover.

And there were some first-rate performances—Ray Charles's gruff, haunting growl through "How Deep is the Ocean?"; Tommy Tune's brisk, exhilarating version of "Puttin' On the Ritz"; Madeline Kahn's squeaky, sexy "You'd Be Surprised"; Tony Bennett's subtle and ominous "Let's Face the Music and Dance" (a genuine existential statement, written long before the distinction was invented); Garrison Keillor's starkly understated reading of the lyrics to "All Alone"; Nell Carter's uproarious romp through "Alexander's Ragtime Band" and Rosemary Clooney's mellifluous "White Christmas."

But there were disappointments along the way, notably Sinatra's under-rehearsed runthroughs of "Always" and "When I Lost You," so sloppily performed that somebody decreed that they be repeated—which they were, to better effect. And the sound in Carnegie Hall—crudely miked and amplified—was abysmal throughout the evening.

Then there was the sad spectacle of Leonard Bernstein. Who else would have had the audacity to sing one of his *own* compositions in a concert devoted to America's greatest songwriter? Bernstein began by playing fragments of Berlin's "Russian Lullaby," interrupting the song several times to tell us how it was one of *his* favorites, and how this was *his* Carnegie Hall singing debut, and a host of other personal irrelevancies. Then he broke into a "12-tone Lullaby," which, aside from its dedication and the repeated invocation of Berlin's name, seemed addressed to Bernstein's own aesthetic concerns, rather than anything to do with the composer of "Alexander's Ragtime Band." "Irving Berlin, I'm sorry!" Bernstein sang, again and again. So, one gathered from the embarrassed tittering in Carnegie Hall, was the audience.

But what was most disappointing about the whole affair was the way Berlin's songs were chopped up and smoothed out for the infamous thirty-second presumed attention span of the television audience. Irving Berlin deserves a serious examination of his creations, which have beginnings, middles and ends, not just quick, catchy phrases, to be tossed off on the way to another song. There was very little sense of the pure, classical structure of Berlin's best work. Little was presented in complete form, and there was the unmistakable flavor of Vegas to the whole affair. It will probably make for pretty good television, and it earned a lot of money for some good causes, but the evening was in no way an apt representation of what Irving Berlin has given us.

I suspect that we will have to "rediscover" Irving Berlin, in much the same manner that Stephen Foster was "rediscovered" in the early 1970s. Everybody thought that they "knew" "Old Folks at Home," "Jeannie with the Light Brown Hair" and the other Foster standards, but the songs had been so distorted over the years—yoked into medleys, decorated with false harmonies, and the like—that it became almost impossible to hear them fresh, for the minor masterpieces that they are. And then Nonesuch Records released an album of Stephen Foster songs, played on an old piano, from the original sheet music, in all of their unsullied glory. And it was a revelation.

On the other hand, the way which Berlin survives such processing is to his eternal glory, and one doubts that anybody could have watched Wednesday night's concert unmoved. There's no business like show business, Berlin wrote, and Wednesday night was show business. But Berlin's songs are more than that.

Newsday (1988)

One does not necessarily improve a sonnet by puffing it out into an epic. Duke Ellington's jazz compositions and popular songs are rightly numbered among the glories of American music, but Ellington had no more business working with symphonic forces than Irving Berlin or Cole Porter, both of whom wisely avoided the temptation and stuck to what they did best.

Sunday night, the American Composers Orchestra proved this handily with a concert devoted entirely to Ellington's symphonic works at Carnegie Hall under the direction of Maurice Peress, augmented by the expert sax playing of Jimmy Heath and the idiomatic pianism of Rolland Hanna. The performances were both loving and spirited, but these compositions simply do not *work*, and I have difficulty believing that anybody would have considered them worth excavating, had it not been for Ellington's other accomplishments.

The four works that were played—a late piece called "Les Trois Rois Noirs," "Black, Brown and Beige," "New World A-Comin'" (a piano concerto of sorts) and a symphonic poem called "Harlem"—are unfocused and meandering. One section follows another, and then another, another and another, until the music suddenly stops (usually after getting as loud as possible and then just a little bit louder). Not only does Ellington write "classical" music like a talented novice; in these works, he serves up disturbingly tame, second-hand, polite and diffident jazz.

In his symphonic works, Duke Ellington, one of the natural masters of jazz, writes it like a tourist. It is as if he had somehow bought the old lie that American music should emulate European classical tradition, rather than striking out on its own path. Everything is careful, correct, bloated and very, very serious—from the pseudo-Rachmaninoff (with impressionist implications) of "New World A-Comin'" to the self-conscious formalism of "Black, Beige and Brown."

It's a pity. A great popular song is a much more useful, and "artistic," thing to have around than an uninteresting symphony, and a composition for jazz band is not necessarily improved by arranging it for orchestra.

Duke Ellington's best work has irresistible energy, originality and no small degree of sophistication—his *own* sophistication, not mannerisms borrowed from Ravel, Debussy, Stravinsky and MGM. And there is more of the divine in the trap drum part for "Take the 'A' Train" (written by Billy Strayhorn but immortalized by Ellington's band) than in all of "Black, Brown and Beige."

Newsday (1988)

19 THE KRONOS QUARTET

One wants very much to like the Kronos Quartet. This San Francisco-based group has devoted its energies entirely to 20th-century music, commissioning and playing works in a panoply of styles. Such catholicity is admirable and this listener has been more than happy to put up with some aggressively trendy programming (as well as the studied grooviness of the group's image) for the genuine musical rewards the Kronos Quartet has offered over the course of its unusual career.

But something was seriously amiss at Alice Tully Hall on Saturday night. Much of the playing was sloppy and thoughtless. Most crucially, instead of meeting different aesthetics on their own terms, the Kronos musicians tended to streamline and homogenize whatever they played into an all-purpose modernism. It may be that the group is simply trying to do too much—too many composers, too many styles, too many performances throughout the country, throughout the world. Success came suddenly, with furious intensity, to the Kronos Quartet, and I suspect that the players may be in a period of transition.

Four works were played, all of them written in the past six years. The "war horse" was Alfred Schnittke's String Quartet No. 3, from 1983. Schnittke is probably the best-known officially sanctioned avant-garde composer in the Soviet Union. I admire much of his work, but found little to enjoy in this strange pastiche of quotations from Orlando di Lasso and Beethoven, padded out with filler that sounded like scraps from Shostakovich's notebooks. It seemed overly referential, a series of musical footnotes, dependent on the listener's background and musical education to provide a context.

Tina Davidson's "Cassandra Sings" was more interesting. This is a solemn, ambitious work that begins with a wailing cadenza for cello and gradually builds into a knotty complexity. Eventually, all devolves into chaos, and it was during this section that the work seemed most clichéd and ordinary. But the spare, beautiful chorale-like melody that followed made for an effective emotional climax. Davidson is a serious composer and one will follow her career with more than usual interest. The Polish composer Henryk Mikolaj Górecki's Symphony No. 3 is one of the masterpieces of our time. His string quartet, "Already It Is Dusk," written for the Kronos, is not quite that but should prove an attractive addition to the repertory—alternately hushed and howling, chantlike and chromatic, often stirring.

The program closed with Terry Riley's "Conquest of the War Demons," a long movement from a longer work, "Salome Dance for Peace, Part Two" that, according to the notes, "follows Salome through a series of dances and confrontations in the underworld and on earth as she pursues a successful quest for world peace." Despite the grandeur of its conception, the quartet, proved skipping, nattering, simple-minded kitsch, indulgent and disorganized in the extreme and wildly overlong for what it had to say. The decline and fall of Terry Riley is one of the saddest stories in contemporary music.

Newsday (1989)

20 GEORGE ANTHEIL REVIVED

George Antheil (1900–1959) styled himself "the Bad Boy of Music," but he was really pretty harmless. His aesthetic has had virtually no influence on anybody; his "innovations" now seem little more than clever gimmickry, and his music has demonstrated precious little staying power. Antheil is likely to be remembered as one of those characters remarkable for the people in his circle rather than for any personal accomplishments—an engaging marginal figure along the lines of Nathalie Barney, Dorothy Brett, Harry Crosby or Alexander Woollcott.

There is no doubt that Antheil was associated with some extraordinary people: His friends included James Joyce, Pablo Picasso, Man Ray and Ezra Pound. Indeed, Pound even wrote a book extolling Antheil's music (thus proving that the poet's ear was as faulty as his political judgment). But Antheil's principal fascination is extramusical. His early biography, in particular, is of uncommon interest: A boy from Trenton, New Jersey, goes to Paris and becomes the darling of the avant-grade, returns to America five years later with a "ballet" for eight pianists, eight percussionists playing xylophones and bass drums, four people responsible for such extramusical effects as sirens, wind machines and propellers and, finally, a mechanical instrument called the pianola.

Unfortunately, the "Ballet Mécanique," described above—and played for the first time in sixty-two years in its original version at Carnegie Hall on Wednesday night—is one of those works of art that are more fun to *hear about* than they are to actually *hear.* The concept is an engaging one, and Antheil's music must have sounded shocking to unprepared audiences in 1927 (although the Italian Futurists and the English Vorticists had fairly well shown the way). Still, the ballet, a Busby Berkeley fantasy on paper, becomes fodder for an Excedrin headache in performance. The initial effect is arresting—in much the same manner as a volley of firecrackers—but Antheil does not have the skill to sustain the work for five minutes, let alone for the half-hour that the performance reconstructed and directed by Maurice Peress took on Wednesday.

Peress, by the way, assembled a superb cast of players for this re-creation and conducted with energy and persuasion. The program, a near-exact duplication of the one Antheil presented at Carnegie on April 10, 1927, also featured a one-movement Sonata for Violin and Piano (with drums, played by the pianist, in the finale); a String Quartet and a "Jazz Symphony" for Piano Solo and Jazz Band. The Symphony was probably the best selection of the evening—one admired Antheil's sonic evocation of a whirling carousel. But it ultimately seemed little more than a "Rhapsody in Blue" without melodies, which is to say a "Rhapsody in Blue" without reason for being.

Randall Hodgkinson, who has developed into a remarkable musician, and Charles Castleman delivered a brilliantly virtuosic performance of the Sonata No. 2 for Violin and Piano. Ivan Davis' playing in the Symphony was dazzling, and the excellent Mendelssohn String Quartet made about as good a case as can be made for the String Quartet No. 1.

In later years, Antheil's style grew increasingly conservative. Perhaps this should not surprise us. One suspects that Antheil understood only the novelty, the shock value, of modernism, rather than the depths and subtleties of the language that was then evolving in the work of contemporaries such as Igor Stravinsky, Arnold Schoenberg, Aaron Copland and others. The bad boy never grew up.

Newsday (1989)

21 BERNSTEIN CONDUCTS COPLAND

Superlatives are always dangerous but, in this particular case, they are also irresistible: On Friday afternoon, the most celebrated conductor in America's history led the New York Philharmonic in works by America's greatest composer. It's as simple—and as complicated—as that. Coming so soon after the deaths of Virgil Thomson and Irving Berlin, however, the concert was a bittersweet occasion; one was acutely aware of the passing time.

Leonard Bernstein was once a ubiquitous presence in New York— part of the terrain, as it were, taken for granted. Now each appearance is an *event*, something to be eagerly anticipated, savored fiercely, reflected upon with gratitude and satisfaction: The *Wunderkind* has become a grand old man. Aaron Copland, too, was once omnipresent, as a composer, conductor, teacher, aesthetic politician and gentle friend to generations of young musicians; I can see him in my mind's eye as I knew him two decades ago, loosely knit, good-humored and absent-minded, sharing a score with worshipful students in the Tanglewood music library.

Copland, who turns eighty-nine in two weeks, was not well enough to travel into the city to hear his protégé and most eloquent advocate conduct four representative works on Friday. Bernstein came through for him, though, as he almost always does; and the New York Philharmonic played with panoramic brilliance and style. It was sobering, however, for one who has long loved these works, to realize that they are no longer "contemporary music" at all but *classics*, written in response to imperatives that are no longer imperative, reflecting and exalting a world that is rapidly fading to memory. Someday, and not too long from now, the idea of a Copland concert conducted by Leonard Bernstein will tantalize our grandchildren in the same way that we are tantalized by tales of Caruso and Stokowski.

Historic stuff, in other words, fashioned in America with a craft and originality that had hitherto eluded us, authoritatively conducted by the first American conductor to win world renown. And great fun to boot—Copland may be the last important composer of concert music to enjoy both the near-unanimous approval of his colleagues and the love of the general public.

The program began with "Music for the Theater," a charming but relatively unimportant early work with a marked jazz influence. Bernstein, hands on wriggling hips, brought just the right touch of stylized vulgarity to the movement marked "Burlesque." Even at twenty-five, Copland had his own distinctive voice: No other composer could have written the plaintive, transparent introductory passage for strings.

"Connotations," on the other hand, is a lost masterpiece. It was commissioned in 1962 to open what was then called Philharmonic Hall and, by extension, Lincoln Center. Bernstein conducted that opening concert. Now twenty-seven years later, in the completely refashioned Avery Fisher Hall, he led a prismatic and urgently dramatic reading of "Connotations." With its granitic angularity, its hard, arching beauty, its ferocious dissonances that explode into even more unsettling major chords, and the shattering conclusion that couldn't possibly get any louder and then does, this represents Copland at his most challenging, and it has never been a popular work. But it is worth the effort.

Stanley Drucker offered a virtuosic rendition of the sedate and endearing Clarinet Concerto, piping brilliantly over ethereal and appropriately vaporous accompaniment from the strings. The program closed with a rather histrionic but ultimately effective rendition of "El Salon Mexico." Bernstein made a more complicated affair of it than usual, breaking up the line with little pauses and "touches" that occasionally seemed impedimental. But the finale was thrilling, as Bernstein let Copland's snatches of mariachi, violent rhythms and earthquake tympanis hurtle, unfettered, to an apocalyptic close.

Newsday (1989)

A little provincialism can be good for the soul. If I were asked to name two parts of the world where particularly engaging and original music is now being written, I would unhesitatingly select Scandinavia and Eastern Europe. Their remove—geographic, cultural and political— from the European mainstream has enforced unusual creative independence. Composers Aulis Sallinen, Joonas Kokkonen and Einojuhani Rautavaara, in Scandinavia, and György Ligeti, Witold Lutoslawski and Andrzej Panufnik, in Eastern Europe, for all their differences, share a craggy, idiosyncratic individuality, little affected by shifts of aesthetic fashion.

These composers are by no means naive, nor are they primitives. They have absorbed modernist techniques, and they know and understand the great musical literature of other lands. But it is a *transplanted* understanding, and by necessity very different from the results of exposure to an unbroken musical continuum. And seeds grow differently on unfamiliar soil.

Henryk Mikolaj Górecki, born in 1933, has lived most of his life near the Polish city of Katowice; until recently his music was unknown in the United States. However, with the release of Górecki's somber, prayerful Symphony No. 3 (and its incorporation into the Maurice Pialat film *Police*), he is finally receiving the attention he deserves. Indeed, the symphony has been recorded three times in ten years—all but unheard of for a major new work.

The Third Symphony (subtitled the *Symphony of Sorrowful Songs*) was composed in 1976. Scored for soprano and orchestra, it is laid out in three movements, all marked *lento,* the first of them nearly a half-hour in length. While Górecki's earlier music was couched in a fairly traditional modernism—esoteric timbres, fistfuls of dissonance and the rest of the arsenal—the Symphony No. 3 seems to have evolved in equal part from Gregorian chant, Polish folk music, Wagner, Messiaen and the minimalists.

Górecki took his texts from three sources. The first movement, largely instrumental, takes the form of a bell-shaped curve. A canon emerges from a grumble of lower strings and develops insistently until the entire orchestra is singing along. The soprano enters with a brief monody, a setting of a fifteenth-century religious poem. The canon then re-enters, full force, plays itself out, and fades into turbidity. The central movement—the shortest, most operatic and most affecting of the three—takes its text from a graffito scratched by a Polish prisoner on the wall of her cell in a gestapo jail. (She signed it "Helena Wanda Blazusiakowna, 18 years old, in prison since 25 Nov. 1944.") The finale is a set of variations on a folk theme from the Opole region of Poland. It is a mother's lament on the loss of her son, mitigated by her ecstatic vision of a future life: "You little flowers of heaven,/please blossom all around him./So that my poor little son/can sleep happily on."

The three recordings of Górecki's symphony all feature the same soprano, the venerable Stefania Woytowicz. The first to be released—the version that was used in the Pialat film—is by far the poorest: an Erato LP (ERA 9275) issued with a preposterously inappropriate cover, murky sound, no liner notes and a terrible pressing. Between the second and third recordings, both available on compact disc, there is a genuine choice. On Schwann (CD 311 041 H1), Wlodzimierz Kamirski leads the Radio Symphony Orchestra of Berlin in a reading that is direct, propulsive and as dry-eyed as possible: It seems to spring from one impulse, sustained for forty-five minutes (more than seven minutes faster than either of the other two recordings). Woytowicz sounds better here than she has in years. In most moods, I favor this rendition.

But I would not want to be without the Olympia recording (OCD 313), featuring the Polish Radio National Symphony Orchestra of Katowice, conducted by Jerzy Katlewicz. It is a much slower, grander (some would say grandiloquent) reading. Still, there is something deeply moving—and apt—about its steady, inexorable and frankly emotive majesty. Both the Olympia and the Schwann recordings also include a charming piece of filler: Górecki's "Three Pieces in Olden Style." The Olympia disc also contains a short Amen for choir.

It is difficult to be objective about the Górecki Third. Some find it sentimental, even bathetic (composer Steve Reich recently offered a withering assessment in *Fanfare*). I happen to love the symphony. From the grave, mantric, contrapuntal progression of the opening movement through the "Isolde meets Satyagraha" aria at the finale, it is one of the most haunting and original pieces of our time. Certainly it is a piece that can be disliked, but I doubt that it will be easily forgotten.

Wigwag (1989)

The original New Music America festival was crowded, noisy fun. Presented in the puddlewonderful spring of 1979 at a performance space called The Kitchen in Lower Manhattan, New Music New York (as it was called) was closer to a block party than a traditional series of concerts. Most of the featured composers—Steve Reich, Philip Glass, Laurie Anderson and Meredith Monk, among others—lived in the neighborhood and the rest of us were only a quick 60-cent subway ride away.

It's easy to feel nostalgic for that time, and not only because we were all ten years younger. Manhattan was a different place then: Rents were stable, the clubs were cheap and open late, nobody dressed up or down, even the hardships seemed romantic. Thousands of young people had flocked to the city, defying the warnings of parents, teachers, back-to-the-land folksingers and the social scientists who promised us the game was over. New Music New York was a celebration of a shared aesthetic—indeed, to use a terrible 70s word, a shared *lifestyle.* There was an unmistakable sense of having tapped into something new, even if none of us was sure exactly what it was. The atmosphere was welcoming and determinedly informal. The artists passed in a cavalcade of sound; during the intervals, The Kitchen turned on standing fans to break the heat and a visitor reached deep into garbage cans filled with ice and water to pull out cold, clean bottles of beer.

What a difference a decade makes! New Music America kicked off its tenth anniversary season Wednesday night with a gala benefit at the Brooklyn Academy of Music. A few minutes after 7, a BAM official took the stage to introduce, uh, *Bianca Jagger,* the honorary gala chairwoman. Then a representative of the Philip Morris Co., which helped fund the festival, came on to tell us of the good things his company was doing for mankind, to scattered hisses. Mayor Edward Koch sent a proclamation: Nov. 10–18 has been declared "New Music America 10th Anniversary Week." A squadron of police stood guard outside BAM. And when the evening was over there were luxe parties on three floors, complete with champagne and suckling pig.

New Music America is now big business. All sentiment aside, there is much to be said for this. The presence of business guarantees that the musicians will be paid, and often paid well. It helps ensure the continuation of what originally seemed a pretty quixotic venture. And it brings a new audience, both fashionable and influential, to some deserving artists.

As might have been expected, the stars came out for the benefit. The Philip Glass Ensemble played an atypically serene but exquisitely nuanced performance of "Train/Spaceship" from *Einstein on the Beach*. Steve Reich's "Different Trains" was performed by the Kronos Quartet: The startling, aggressive opening measures—a churning of strings accompanied by the taped shrieks of locomotives long departed—are among the best music the composer has ever written.

And then there was Moondog—best known for his extended, eccentric residency on the corner of Sixth Avenue and 53rd Street, where he stood every day during the 60s and early 70s dressed as a Viking, but a composer of distinct originality as well. He was back from Europe for the first time in some years. With its droning, airy textures and strong sense of ritual, Moondog's beautiful "Surf Session," in which he led several members of the Brooklyn Philharmonic Orchestra, makes apt and erudite reference to the music of the Renaissance yet remains clearly contemporary in spirit.

I am at a loss to explain the presence of Jevetta Steele, Bob Telson and Little Village at this benefit: They offered wonderfully bracing MOR gospel, but there was nothing "new" about it—these slick, clever songs could have made playlists of a mid-70s FM radio station. The Brand Nubians, a rap group, inspired nothing but Spenglerian ruminations in this listener. Host Fran Lebowitz acknowledged that she was out of place in new music circles and was devastatingly funny about it; Allen Ginsberg, equally out of place, rambled incoherently; and Laurie Anderson offered a wry reminiscence of the first festival.

New Music America long ago burst the bounds of The Kitchen, and even the grander spaces of BAM cannot accommodate it all. The festival, produced by Yale Evalev, seems to be everywhere this week, with presentations at Merkin Hall, Symphony Space, the Prospect Park Picnic House, La Mama, Roulette, the Experimental Intermedia Foundation and many other venues. A visit—several visits—may be warmly recommended.

Newsday (1989)

24 HOLY BLOOD AND CRESCENT MOON

Cleveland, Ohio

Stewart Copeland was the percussionist for The Police, a rather fine rock band that flourished in the late 1970s and early 80s. All well and good, but he has about as much business writing an opera as Ringo Starr. Nevertheless, the Cleveland Opera presented the world premiere of Copeland's *Holy Blood and Crescent Moon,* an opera in two acts, at the Playhouse Square Theater here Tuesday night. It was a complete and unredeemed disaster; it was also, of course, completely sold out. Somewhere, P.T. Barnum is smiling.

Copeland, in pre-premiere interviews, cited Carl Orff and Richard Wagner as his major influences. This is an old gambit for rock and roll *poseurs:* When defending one's work, talk up ancient traditions and some fancy predecessors, preferably European—remember Patti Smith and her hare-brained allusions to Rimbaud?

In fact, Copeland belongs to another tradition, that of the privileged dilettante who decides one day to become a composer and then finds not only believers but sponsors. Indeed, I dare venture that the name Stewart Copeland will someday rank proudly with those of Richard Nanes, Gordon Getty and Rebekah Harkness in the roster of our wealthiest American composers.

Not much more can be said for him. Copeland's technical knowledge of music is rudimental at best; he has no sense of theater and small gift for setting the English language. The score of *Holy Blood* consists of little more than a succession of seemingly random, repeated figures; it lacks characterization and linear continuity. Copeland's music has nothing to engage the attention, neither melodic invention nor visceral power nor the hypnotic aural clockwork that we find, say, in the operas of Philip Glass. Fortunately, Copeland had the taste to hire a reasonably skillful arranger and orchestrator, who endows the score with whatever coherence it possesses and is nowhere listed in the credits.

The libretto is preposterous twaddle, a *mélange* of crazed Muslims and Christians shrieking for blood, handsome princes, exotic princesses, swordfights, balcony scenes and immortal lines like, "This is very strange/ Your father is not a violent man/ But someone did attack me/

All I need is a solid horse." And, yes, there is a "message" of sorts—
religious fanatacism is bad, boys and girls—tacked on implausibly at
the end, like a misplaced line from Aesop.

The Cleveland Opera gave *Holy Blood and Crescent Moon* as good
a production as I hope it ever gets, with conventional but handsome
Arabian Nights decor and costumes that might be described as a mix
of Shakespearean and Hollywood Turk. Imre Pallo presided over a good,
disciplined ensemble, the Ohio Chamber Orchestra, and some strong
soloists, among them Edward J. Crafts, Gloria Parker, Tom Emlyn
Williams, Jon Garrison, Marla Berg and Charles Karel, who did their
best to elevate Copeland's ineptitude.

I will not presume to judge the motives that inspired the directors
of the Cleveland Opera to choose *Holy Blood and Crescent Moon* as the
company's first world premiere. One is delighted to see any troupe take
on new work; that said, I cannot imagine that there is a university music
department in this country that couldn't have produced a stronger score.
But would Playhouse Square have been filled? And would the press
have come to Cleveland from around the world? These are serious—
indeed, depressing—questions, and I don't want to think too much
about the answers just now.

Newsday (1989)

25 PETER SERKIN PLAYS COMMISSIONS

Peter Serkin, a great pianist at the peak of his powers, has had the
splendid audacity to devote an entire program to new music—and by
"new music" I *don't* mean the 45-year-old Prokofiev sonata that conservatory students are forced to learn before graduation, as a token
gesture to the 20th century.

No, Serkin long ago proved his devotion to the music of his time,
first with the recordings of Messiaen that he made in the early 70s, later
with impeccable readings of pieces by Stefan Wolpe, Toru Takemitsu,

Peter Lieberson and others. Now he has commissioned short works from eleven composers—an entire evening of new music—and he played the cycle's premiere performance at the 92nd Street Y on Saturday night.

Serkin chose his composers carefully; they have rewarded him with miniatures of unusually high quality. The remainder of this review must necessarily take on some characteristics of a list, for none of the pieces should be left out. Indeed, several of Serkin's commissions deserve to enter the repertory and the best of them deserve an article in themselves.

Lieberson's "Breeze of Delight" began the program—a thoughtful, restrained and beautiful work that, with its traditional, arpeggiated bass line and ethereal melody, sounds a little bit like two disparate pieces by Chopin being played at the same time, one with the left hand, one with the right. Oliver Knussen's Variations proved a chiming, clangorous exercise in compression, without a wasted note. Bright Sheng's "My Song" alternates between aggressive, fiercely virtuosic passages and a gentle, fluttering *chinoiserie*. At one point, Serkin was called upon to reach over the keyboard and strum the strings of the piano; this technique usually seems a gratuitous modernist gimmick but Sheng (and Serkin) made it musical.

Tobias Picker provided three interrelated pieces: The outer two shot by quickly and brilliantly, the central movement was serious and sumptuous. Christine Berl's "The Lord of the Dance" was built in four distinct sections, according to the composer's program notes; I could not have drawn borders, but found Berl's discourse, with its stark harmonies and ringing tremolos, constantly engaging. Tison Street's Romance in D major received one of the warmest ovations of the evening, but I thought it treacly and banal, written in a harmonic language that Rachmaninoff would have rejected as hackneyed.

Alexander Goehr's " . . . In Real Time" is something of a process piece, with a strong rhythmic impetus, and a constantly changing musical landscape. Luciano Berio's "Feuerklavier" begins with a long trill that suggests another famous musical depiction of flame, Falla's "Ritual Fire Dance." It is an effective little work, abstracted musical satellites flying off from a single, insistently repeated note.

Leon Kirchner's Interlude may be the richest piece in the series: It is knotty, complicated yet always directly expressive. Toru Takemitsu's "Les Yeux Clos II" is Japanese impressionism: featuring an inventive ending that catches the listener by surprise but then settles in, ultimately seeming not only logical but almost inevitable. The program closed with a second work by Lieberson, a hearty Scherzo, and an "encore" by

Hans Werner Henze, a compact, flashing little miniature, here and gone.

Throughout the evening, Serkin's interpretations were marked by the intelligence, dignity, musicianship and technical mastery that he brings to everything he plays. Those sentimentalists who like to whine about a lost "golden age" of pianism are simply looking in the wrong places.

Newsday (1989)

26 VLADIMIR FELTSMAN

Vladimir Feltsman first came to world attention as the hero in what one may hope was among the last of Cold War dramas. In 1979, Feltsman, a highly regarded Soviet pianist with a flourishing career, made application to immigrate to Israel. Permission was denied and, for the next eight years, his musical activities were strictly curtailed. A "non-person" in the Soviet Union, Feltsman became something of a legend elsewhere—the brilliant young pianist nobody had heard.

Finally, through a combination of international pressure and a relaxation of government strictures, Feltsman succeeded in procuring a visa in the summer of 1987. He came directly to live in the United States and played his American debut at the White House, following up with sold-out concerts at Carnegie Hall and the Kennedy Center. Monday night, he returned to Carnegie for a recital of works by Schubert, Liszt and the contemporary Russian composer Nikolai Karetnikov.

Feltsman's odyssey was—and remains—a great story. Moreover, by all accounts, he suffered his travails with dignity and forebearance. But many musicians and some critics reacted with a certain skepticism to the claims that were being made for him when he arrived in this country. Nobody could have lived up to this barrage of publicity, and Feltsman's generally excellent debut concert necessarily came as something of a disappointment: I remember thinking at the time that, yes, he was very good but that he was no Vladimir Feltsman.

Since then, however, Feltsman has proven himself. He is an extremely interesting musician and probably the most original and distinctive pianist to emerge from the Soviet Union since Sviatoslav Richter. But he is something of a wild card, with virtually nothing in common with the "Soviet school"—that peculiarly feverish, glassy toned, hyperemotional and ultravirtuosic brand of athleticism, exemplified at its best and worst by the late Vladimir Horowitz.

On the contrary, Feltsman is a cerebral artist, even something of a Modernist. His interpretations are formal, reserved, deeply calculated, meticulously wrought and, by some standards, rather eccentric. Instead of smoothing over the disjunctions in Schubert's Sonata in B-flat and Liszt's Sonata in B minor, for example, Feltsman tended to *emphasize* them (the mysterious rumbling trills in the first movement of the Schubert seemed to emanate from the Carnegie walls).

And yet Feltsman ran the Schubert together as an unbroken utterance, with only the briefest of pauses between movements: The Scherzo evolved from the afterglow of the Andante, while the finale skittered nervously out of the Scherzo. One particularly admired the elegance and clarity of his voicing, the unusual but effective even, almost motoric quality he brought to accompanimental material in the left hand.

Karetnikov's set of "Two Pieces" (Op. 25) was presented immediately after intermission, a U.S. premiere. The first of these proved a competent invention in the international 12-tone style and could have been written by any of a thousand academic composers from Ashtabula to Zanzibar; the second piece, however, is dark, serious, sumptuous and beautiful, and Feltsman made a persuasive case for it.

The program closed with Liszt's Sonata in B minor. The musical world can be divided into two categories—those who believe Liszt was a great composer and those who wonder why. I am of the latter persuasion, but Feltsman's coolly analytical reading of the sonata, which stressed Liszt's structural and harmonic innovations rather than his saccharine melodic modules and fuzzy metaphysics, almost made a momentary convert. Almost.

Newsday (1990)

Arvo Pärt was a prominent film composer in the Soviet Union before emigrating to the West in the mid-1970s. After his repatriation, he began writing spare, austere, predominantly consonant music that combined minimalism with medievalism. His first recording for ECM, "Tabula Rasa," won immediate and deserved acclaim; here was music that could have been written in almost any century after the fifteenth yet seemed absolutely of our time.

Pärt's "Passio" (subtitled "The Passion of Our Lord Jesus Christ According to St. John"), which received its New York premiere at Alice Tully Hall Wednesday night, is a work of high seriousness and if, in the final analysis, I respect it more than I love it, the respect is genuine and grateful.

The composer has fashioned a word-for-word setting of the Vulgate Latin text of the New Testament Gospel of John, Chapters 18:1 through 19:30. "Passio" is scored for solo bass (Jesus); solo tenor (Pilate); a quartet of soprano, countertenor, tenor and baritone (who pass the narrative of the Evangelist among them); a chamber choir; violin, cello, oboe, bassoon and organ.

Jean Sibelius once said of his fourth symphony that there was "nothing, absolutely nothing of the circus in it"; the same observation might be made of "Passio." Its music is chaste, starkly simple, determinedly antitheatrical. The last quality sets it apart from most forerunners in this genre: Even the greatest Passion settings have sometimes resembled vast, metaphysical courtroom dramas. Pärt will have none of that. "Passio" is closer to religious service than to music theater. It makes no effort to enhance the story, no appeal to the adrenal glands; there is, for example, only one forte passage in the entire composition. Indeed, it is among the least manipulative music I know.

And so this "Passio" is stern, dignified, formal and—when it is not beautiful—somewhat dull. There is little substantial difference between the beauty and the boredom: For better and for worse, "Passio" is remarkably all of a piece, but one's reactions to it evolve over its seventy-minute duration, largely because the work itself does not. At times it seems an extended Renaissance madrigal, something that might have been composed by a denatured Gesualdo four hundred years ago, peppered with some Stravinskian dissonances which are usually allocated to the organ and chamber choir.

Pärt's music, with its subtle, static rituals, is very pure, very specific. He goes his own way and I admire his single-mindedness. Moreover, he attracted an unusually distinguished audience to Alice Tully Hall: One couldn't swing the proverbial cat without hitting a composer or music professional. But I confess to battling a recurrent impatience Wednesday night. Virtually nothing "happens": Pärt admits the heavens but not the world. And there is a human drama inherent in the story of the Crucifixion that I find lacking in "Passio," for all of its nobility and elevation.

Still, whatever its flaws, "Passio" is an important work, and its strongest moments are imbued with a gentle, rarefied power. The performance, by the way, was magnificent—committed, disciplined, impeccably wrought: The players fulfilled their duties with the rapt intensity of pilgrims.

Newsday (1990)

28 POGORELICH AT CARNEGIE HALL

Ivo Pogorelich's Monday night piano recital at Carnegie Hall brought out the mad scientist in me. Had it been possible to replace Pogorelich's brain with that of an ordinary first-year conservatory student while leaving his spectacular technical skills intact, we might have heard a better concert. Instead, I found myself reflecting upon the *New Grove Dictionary of Music and Musicians'* famous, idiosyncratic put-down of Vladimir Horowitz (whom Pogorelich in no way resembles)—"possession of an astounding instrumental gift carries no guarantee about musical understanding."

Competition winners are a dime a dozen, but Pogorelich came to attention by *losing* the 1980 International Chopin Competition in Warsaw. Pianist Martha Argerich resigned from the jury in protest, declaring Pogorelich a "genius." The press picked it up, and a star was made.

I don't know whether Argerich still considers Pogorelich a genius—passions run high during competitions—but he obviously likes himself very much indeed. Certainly more than he does Haydn, Brahms, Chopin or any of the other composers that he remade in his own image. Certainly more than he does his audience, which he blithely ignored during most of the ovations by sitting at the keyboard, staring off to the side, the beginnings of a pout on his lower lip.

Pogorelich is a pianist of the "Make it Meaningful" school. He dawdles over *everything* and seems to equate slow tempos with profundity. He willingly sacrifices the big picture for microscopic detail; he not only doesn't see the forest for the trees but doesn't see the trees for the moss on their bark. Subjective to the point of solipsism, Pogorelich has a whim of iron: Whenever he wants to, he slows down; whenever he wants to, he speeds up. Color, timbre and velocity are everything; of rhythm he seems to know nothing at all.

Most of the opening Haydn Sonata in D (Hob. XVI:19) was played very softly, without any central pulse, in tiny little pecks and slaps that were then smeared by overpedaling. Haydn is among the most healthy and extroverted of composers; Pogorelich's performance was pallid and involuted, a fundamental misreading that can only be understood as an arbitrary exercise. Three Brahms intermezzos were not much better: the slowest performances I've ever heard, all hesitation and bejeweled pianissimos—New Age Brahms, beautiful but dumb. Six Scarlatti sonatas were not belabored so much as one had feared but still seemed overfreighted and charmless, and I regretted the decision not to take the repeats.

Just as one's respect for Pogorelich is likely to decrease in proportion to one's familiarity with the music he plays, I have found that the pianist is at his best in second-rate pieces. It is impossible to fiddle too much with Liszt's Transcendental Etudes, and the three presented Monday night received exciting, hypervirtuosic readings. Scriabin's Sonata No. 4 was skittish and loud but often brilliantly effective; Pogorelich has "chops," no doubt about it. He closed the program with Balakirev's "Islamey"—thousands of notes, most of them unnecessary, a display piece so vulgar, glittering, pretentious and altogether mindless that one half-welcomed Pogorelich's "touches."

Newsday (1990)

29 A MOMENT OF SILENCE

André Previn and the Los Angeles Philharmonic brought the best program I've heard from either of them to Avery Fisher Hall Friday night. Still, the evening's most remarkable moment took place not when the orchestra was playing, but during a pause between the first and second movements of the Shostakovich Fourth Symphony.

The Shostakovich is not an easy work but amply repays any concentration that a listener may bring to it. After the long first movement, however, there was the usual noisy exodus by disgruntled patrons, apparently upset by the introduction of dread dissonance into their digestive meanderings.

What followed was an aesthetic Boston Tea Party of sorts. As the escapees clacked righteously up the aisle, Previn gently, effectively fought back. He turned and regarded the disruption with a gaze of weary amazement, shaking his head. The rest of the audience caught on and began to clap and stomp—at first facetiously, in honor of the departing guests (who wisely quickened their step) and then with genuine appreciation for Previn, the orchestra and, especially, Shostakovich. When the symphony ended, a noisy, harrowing and ultimately ennobling half-hour later, the listeners who stuck it out—about 90 percent of the hall, by my count—rose and provided an ovation of unusual intensity.

More than half a century ago, composer Charles Ives, enraged by the rejection of a new work, challenged the recalcitrant audience to stand up and use its ears "like a man." Discounting the dubious gender reference, what excited me about Friday's concert was the willingness of the Lincoln Center audience to do just that. Those who did not leave after the first movement were galvanized into acceptance, as a point of pride. They seemed to understand that Shostakovich was not writing difficult music to insult or confuse them, nor to play an obscure theoretical game, but rather to tell the truth about his life and times, in a language beyond words.

Now, when immediate "accessibility" is the highest aspiration of many contemporary composers and the two commercial classical radio stations in New York have effectively banned most 20th-century music, it is heartening to witness a demonstration such as this. Perhaps, one dares to hope, the orchestral repertory has a future and not just a glorious past. In any event, the Shostakovich symphony—written in 1936, during one of the darkest years of Stalinist terror, and suppressed by the composer (who feared for his career, if not his life) until 1961—was brilliantly played by the Los Angeles Philharmonic: spacious and centered, virtuosic and purposeful.

The program began with Beethoven's ingratiating B-flat Symphony, also its composer's fourth. Here, Previn and the orchestra were less successful. The opening Adagio seemed oddly timid and disjunct, restrained rather than merely slow, and the sudden acceleration into Allegro Vivace did not follow with the inevitable logic that other interpreters have found in this music. Throughout the symphony, the playing was rather heavy and Previn's interpretation, by and large, lacked charm. Still, I admired the steady, formal Haydn-like qualities that he brought to the Scherzo and there was some fine, chugging, rushing string playing in the finale, a Jovian romp.

Newsday (1990)

30 OPENING NIGHT AT THE MET

Suddenly it is autumn, and New York is reborn. The evenings are cool; the streets are lively; the restaurants hum with bright, eager conversation. The gaseous miasma of Manhattan summer is past, and the collapse of Western Civilization (believed imminent throughout July and August) seems, once again, a safe distance away.

Although New York is no longer without music in the summer—Mostly Mozart and the New York City Opera season have radically changed the migratory habits of our performing artists—September is the month when simmer comes to boil. The New York Philharmonic is back in place at Avery Fisher Hall; Carnegie Hall begins its centennial season tonight. And, on Monday, a distinguished audience made its way to the Metropolitan Opera House, to celebrate the opening night of the 1990–91 season by dressing up in fabulously impractical evening wear, sipping and socializing through vast intermissions and—not-so-incidentally—experiencing once again the thrice-familiar tale of hungry, unbowed Rodolfo and his gentle, tu-bercular seamstress, Mimi.

Opening night is the priciest ticket of the Met season, and there is a certain irony inherent in the spectacle of penguined and de-la-Renta'd power brokers shedding tears over the fate of starving artists. But *La Bohème* is a difficult opera to resist; I suspect that anyone who was ever young in a big city and devoted to an art likes to think it was something like this—the late nights in crowded cafes, the fast-and-easy anti-authoritarianism, the idyllic cheapness of young love.

Indeed, the *Bohème* characters are timeless: There are Rodolfos on the subway, Mimis on St. Mark's Place, Collines in the upper reaches of Morningside Heights. Moreover, the score, while fitful and episodic in the first, second and fourth acts, contains some of Puccini's most affecting melodies; the third act, a gorgeous, unbroken continuum, may be the composer's most perfect creation.

Opening nights do not always represent the Met at its finest (su-perstars jet into town, attend some quick rehearsals and then move slowly about the stage like gods and goddesses trapped in amber) but this *La Bohème* was an exception, and credit must go to David Kneuss for his remarkably convincing and unified stage direction.

Placido Domingo sang Rodolfo's softer passages with a honeyed sweetness; in heroic passages, his voice darkened noticeably, and he occasionally sounded strained. Still, one admired his ardor and dra-matic intelligence, rare for a tenor. Mirella Freni was a touching and idiomatic Mimi. The fragile, passive—and altogether winning—charm that has been her stock in trade for thirty years shows no signs of depletion, despite an occasional wobbly tone and some ob-vious anxiety preparing the very highest notes of "Mi chiamano Mimi."

Brian Schexnayder was a dignified, somewhat stolid Marcello, while Barbara Daniels brought animal spirits and a healthy vulgarity to the role of Musetta. Julien Robbins, as Colline, sang the "Coat Song" with a welcome directness; he refused to milk its pathos, making it all the more effective. Richard Cowan, as Schaunard in his Met debut, proved a genuine singing actor; for once, the character seemed flesh and blood, a vital part of the action, rather than merely the "other" Bohemian. Christian Badea's sure, sensitive conducting was marked by an unusual feel for the score's ebb and flow, despite a recurrent tendency to let the orchestra overpower the singers.

Finally, there is Franco Zeffirelli's production. The first and last act are overblown and distancing: One views the protagonists as through a telescope, backwards. But the second act seems to me Zeffirelli's masterpiece—an evocation of nineteenth-century Parisian city life so vivid and multifaceted that one is half-tempted to saunter up onto the stage and order an absinthe.

Newsday (1990)

31 PETER SELLARS' FIGARO

Boston, Mass.

Peter Sellars, the creative, controversial young director who began his maverick career while still a student at Harvard, came home last week to a hero's welcome.

Everyone east of Worcester, it seemed, wanted a ticket to the opening production of Sellars' new Boston Opera Theater—the latest effort in the long and seemingly Sisyphean struggle to establish a regular opera series in this most proudly cultivated of American cities. Those who made it past the ornate lobby of the Colonial Theater—all five performances sold out immediately—were rewarded by a *Marriage of Figaro* that was as tender, touching and true as any I've seen, with superbly integrated ensemble work from Sellars' singing actors and a ingenious directorial concept that, for once, seemed absolutely faithful to the spirit (if not the letter) of Mozart and Da Ponte's intentions.

By now, Sellars has become a critical litmus test of sorts; the director has been exalted and derided with unusual ferocity—sometimes within the pages of the same journal. Throughout it all, I have sat uneasily on the fence. Sellars' *Don Giovanni,* for example, struck me as a desecration—arty, contrived and false to its source. The action was set in an American city slum; the men were junkies and pimps, the women were whores. The whole question of moral distinction and deliberation, around which the opera resolves, was effectively obfuscated.

But this *Marriage of Figaro* is different. In the past, one could almost imagine Sellars' thought processes as the action progressed ("Gee, wouldn't it be *cool* if...?") However, in *Figaro,* he has put the hyperactive cleverness that has characterized his work from the beginning to the service of the opera at hand. How perfectly apt, for example, to render Cherubino a big, clumsy child in a hockey jersey, who never walks when he can leap, and alternately undulates his hips and checks the fridge during "Non so piu"—a real adolescent, in short; someone we *know* rather than a corseted diva in a Robespierre wig.

There were plenty of such "touches" throughout the evening. The action is set in Trump Tower, where the modern equivalents of Da Ponte's counts and countesses now dwell. Figaro and Susanna are a genteel, modern-day butler and maid, their wedding bed a fold-out couch in the laundry room. The Count, decked out in high fashion, commands a gigantic glassy living room, surrounded by vistas of steel and smoke. The wedding is dutifully recorded by a VCR, and the guests shimmy and shake through a sinuous break-dance divertimento played on a compact disc.

Sellars and his adept, sympathetic conductor, Craig Smith, observed extremely long pauses during recitatives—a radical idea that worked brilliantly. And so, instead of the incessant, keyboard-accompanied chatter that we usually find in these passages of dialogue, the characters took their time with their words, as if they were actually mulling over what they were saying, rather than just rattling off melodious Italian.

The performances were lively, intelligent, heartfelt, interactive and eminently believable; one had the sense that the characters all lived in the same world. Sanford Sylvan's Figaro was gentle and vulnerable, even depressive in Sellars' hauntingly sad representation of the final act. Jeanne Ommerle sang Susanna brightly and with dulcet grace, while Jayne West brought patrician sorrow and infinite forgiveness to the Countess. James Maddalena was convincingly unctuous as the arrogant Count, although I wish the scenes of domestic violence could have been toned down a degree; it is hard to feel any sympathy for a wife-beater.

Susan Larson brought a detailed and wonderfully awkward ardor to the role of Cherubino. Frank Kelley played Basilio as an oily, leather-clad rock musician, while Lynn Torgove rendered Barbarina a squeaky, sexy little tart in a tight skirt. Sellars threw nothing away; even the smaller roles were vividly detailed. Sue Ellen Kuzma's Marcellina began as a cronelike caricature and evolved into a warmly empathic, fully drawn character and David Evitts made a genuinely interesting figure out of Bartolo. Herman Hildebrand blustered effectively if perhaps too effusively as Antonio, here ensconced as the superintendent of Trump Tower.

Recounting these details emphasizes the humor and strangeness of Sellars' *Figaro*, which does both director and production a disservice. On the contrary, this is, at bottom, a profoundly serious staging of a masterpiece. The final scene takes place on the balcony of the 52nd floor, amidst those strange, transplanted fir trees so endemic to wealthy terraces, as the characters, far from the Earth and shut out of the apartment's warmth, prance and fumble through an elaborate and hurtful charade. Finally, at the edge of this man-made precipice, through the efforts of a single character, there is forgiveness and reconciliation. Rarely has it seemed so wondrous, so healing, so desperately needed.

Newsday (1991)

32 MITSUKO UCHIDA'S MOZART

Mitsuko Uchida plays Mozart with clarity, style, virtuosity and a welcome element of ferocity. Her Friday night concert at Alice Tully Hall, devoted to four early sonatas, will likely be remembered as one of the highlights of Lincoln Center's ongoing bicentennial salute to the composer. Happily, there will be a sequel this Sunday at 3, again at Tully Hall, when Uchida will play four more sonatas and the Fantasia in D minor (K. 397). Anybody with a love for Mozart—or, indeed, for top-rank pianism—should be there.

The first and foremost thing that impresses a listener about Uchida's Mozart is its sheer virility. Gone—banished for the duration—is the pretty, periwigged *Wunderkind* of antiquated legend. Uchida makes no attempt to emulate a music-box; she obviously wants these sonatas to do more than tinkle and charm. And, in her hands, they sound passionate, full-blooded, charged with drama, almost epical.

The root of her artistry lies, I think, in her sense of steady rhythmic propulsion. This is not to suggest that she forces any sort of straitjacket on the music, only that one generally has a sense that she knows just where she is going. She does not dawdle, she does not sentimentalize; her playing, particularly in faster movements, is goal-oriented, in the complimentary sense of the phrase.

Uchida is capable of spinning out a sweet, songful andante with the best of them, but even then her interpretations are characterized by what is, for this music, an unusual emotional boldness. An earlier generation, which prized repose and balance as the highest Mozartean virtues, would likely have found her playing rather shocking. I love Uchida's work; even her occasional recklessness seems bracing after so many years of rapt and studied politeness in Mozart performance.

Indeed, my one complaint would be with her trills, which often sound self-conscious and artificial—as if they were grafted onto the music as an afterthought. They are almost *too* even, *too* pearly, *too* carefully prepared for the general dynamism of her conceptions. For a moment, one's attention is drawn to Uchida's stylistic knowledge and technical command and away from the music itself (this was particularly noticeable in the Adagio from the Sonata in F, K. 280). But such moments were rare on Friday night.

Newsday (1991)

33 THE DEATH OF KLINGHOFFER

Two weeks back and some two miles away from the Brooklyn Academy of Music, where *The Death of Klinghoffer* received its American premiere Thursday night, a car driven by a Hasidic man careened out of control and killed an African-American child. Some area residents (led on by visiting "activists") used the tragedy as an excuse to run riot in the streets—destroying stores, beating up photographers, reporters and any perceived enemy in sight. A young Hasidic student unrelated to the driver, whose only "crime" was his religious faith, was fatally stabbed.

If John Adams and Alice Goodman ever decide to collaborate on an opera about the recent unpleasantness in Crown Heights, we may be sure that all parties involved will be treated with equal courtesy. Or so one would gather from *The Death of Klinghoffer*, a pompous, turgid, derivative and hopelessly confused operatic meditation on the 1985 hijacking of the cruise ship *Achille Lauro* and the resultant slaughter of an elderly, wheelchair-bound passenger, Leon Klinghoffer.

To be sure, the Palestinian terrorists, as Goodman presents them, get a little carried away now and then but, after all, they are Redressing Ancient Wrongs. They have a *purpose* to their lives; they are *real men*—Rousseau's "Noble Savages" made flesh—as opposed to the opera's nattering, ineffectual Jews, with their bank accounts, their Evian water, their bad bowels, their collections of souvenir kitsch, and their use of funny words like "meshugaas."

At least it was Adams and not Goodman who spoke in interviews of the opera as a "healing thing," for there is nothing ameliorative about the libretto. I don't think that it is actively anti-Semitic, as some have charged; instead, Goodman's words drip with contempt for the creature comforts that (one may guess) most of the affluent audience that filled BAM on Thursday night enjoy as a matter of course. Add to this the requisite fascination with the "Third World"—"We are not vandals but men of ideals" one of the hijackers declaims—and you have the Politically Correct opera *par excellence*: unreflected, privileged, simultaneously apologetic and mischievous, as trendy as spiked hair. Despite the occasional majesty of Goodman's phraseology, I was ultimately reminded of nothing so much as one of V.S. Naipaul's parlor revolutionaries, traveling off to distant lands and preaching the gospel, all the time fingering that crucial return plane ticket in the pocket.

The music, on the other hand, may perhaps be described as "healing"—if one happens to be soothed by warmed-over Vaughan Williams, spiced as necessary with a gentle chromaticism and gestures appropriated from composers as diverse as Puccini and Philip Glass (even Ponchielli seems to dance by, hourly). Adams is greatly admired in some circles but, in all good faith, I could hear nothing that was uniquely *his* in *The Death of Klinghoffer*, except a marked improvement in his orchestration since *Nixon in China* and a rather increased understanding of the human voice. Whatever one thinks of the Glass operas, they are at least something tangible that one may take or leave, according to taste. There is no such specificity in *The Death of Klinghoffer*; instead, Adams offers a bland compote of minimalism and the Western classical tradition—new music that is not new; old music made *chic*.

Whatever pleasures the evening afforded lay in the customarily creative staging by Peter Sellars and the splendidly unified performances by the members of his ensemble, some of whom played several roles (Sanford Sylvan was not only a tender Klinghoffer—he endowed the character, apparently designed as little more than a "specimen," with a haunting poignancy—but Klinghoffer's dyspeptic neighbor Harry Rumor as well, while Janice Felty essayed a Swiss Grandmother, an Austrian Woman and a British Dancing Girl). Sellars' use of videocameras had the dual effect of allowing us to witness, close-up, the faces of the fanatical and the frightened (what a wonderfully expressive face Felty has!) while distancing us from the characters in the act of throwing them on a screen—a dichotomy the director surely intended.

There was visceral, idiomatic support from Stephanie Friedman, Thomas Young, Thomas Hammons and Eugene Perry. Kent Nagano's conducting was seemingly authoritative and appropriately adaptive. James Maddalena played the Captain with a palpable sense of shock while Sheila Nadler strove nobly to convey the sort of cookie-cutter Mahler-meets-Sophocles-in-Suburbia grief Goodman and Adams have attributed to Marilyn Klinghoffer.

Newsday (1991)

34 ERNST KRENEK AND JOHN CORIGLIANO

This space was originally reserved for a celebration of the Metropolitan Opera's new production of *The Ghosts of Versailles,* and, especially, of the fact that the composer and librettist, John Corigliano and William M. Hoffman, were greeted with a roaring, stomping and altogether unbridled standing ovation at the conclusion of the first performance.

The death of Ernst Krenek, last week in Palm Springs, California, at the age of 91, changed my plans—but not entirely, for there is a connection between Krenek and Corigliano. Indeed, they had something crucial and distinctly unusual in common: They were among the only composers who lived into the last decade of our century who were honored with full productions of their operas at the Met.

"There was an imposing audience and the applause signified a deep satisfaction," W. J. Henderson of the New York *Sun* (probably the city's best music critic in the dreary days between the death of James Huneker and the advent of Virgil Thomson) wrote of the January 10, 1929, American premiere of Krenek's *Jonny Spielt Auf!* "The grand opera music is frequently dissonant in the approved style of the day and there are some peppery instrumental combinations," Henderson continued. "The xylophone, the rattle, the railroad whistle and the automobile horn lend their divine aid."

Particularly famous was the climactic scene, which featured the opera's protagonist, a jazz saxophonist, played by baritone Lawrence Tibbett, in blackface, astride the world, Nero's fiddle in hand, with the skyscrapers of New York (most notably the Woolworth and Singer buildings) towering behind. It was a few months before the stock market crashed, the end of the "jazz age," and Krenek's mixture of vernacular styles and high modernism must have seemed the apogee of the avantgarde.

There has never, to my knowledge, been a complete recording of *Jonny Spielt Auf!*" A great deal of Krenek's work has never been recorded, including the opera he considered his masterpiece, *Karl V* (1930–33). And, as one of Krenek's most ardent and articulate champions, the late Glenn Gould, observed, it is through the media—recording, radio, television, film—that one is most likely to encounter music these days.

So *Jonny Spielt Auf!* has remained something of a mystery. There is much that one would like to know about it: How did the austere Krenek, who was temperamentally uncomfortable with most jazz and popular music, manage to incorporate these elements into his work? What are the similarities between *Jonny* and Weill's *Threepenny Opera*, both of which come from a similar time and place and reflect a not-dissimilar world-view? What are the qualities that so scandalized audiences in Leipzig and at the Met? Has *Jonny*—like Strauss's *Elektra* but unlike *Pierrot Lunaire*, *L'Histoire du Soldat* and many other pivotal modernist works—retained the power to shock? Is it as expressive—as genuinely *beautiful*—as some of Krenek's string quartets and "In Memoriam Anton Webern"?

It is in many ways ironic that the composer of *Jonny* should die the same week that *Versailles* received its premiere, for Krenek and Corigliano, while both excellent composers, were radically different in their chosen aesthetics. Krenek's musical utterance was generally sober, determinedly unified in its syntax and proud of its uncorrupted "modernity." Corigliano's work, on the other hand, is playful, eclectic and wildly allusive, its abundant modernism sweetened with a host of elixirs.

Which composer is "right"? Is Corigliano's attempt to reach a large audience pandering, as some of his elders (and more than a few envious "youngers") might claim? Was Krenek guilty of an ivory-tower elitism? As one who loves music by both men, I am perhaps not the one to judge. Indeed, as I grow older, I am both more and more convinced of the importance, the inherent *seriousness* of life and less and less satisfied by proffered answers to its riddles—by any version of the proverbial One True Faith, whether political, religious or creative. (Fundamentalists, Marxists and capital-M Modernists have a lot more in common than they might like to admit.)

And so I reject the "Oh-Wow-Man" school of new music fandom that would write of complex masterpieces by Anton Webern, Luigi Dallapicolla, Elliott Carter—and Ernst Krenek—as elitist and desiccated with the same heat that I reject the insular, academic putdowns of composers like Philip Glass and Corigliano. There is no need to arbitrarily restrict our diet. I've written this article listening to *Phil Spector's Christmas Album*; on another occasion, in another mood, I might well have written about Phil Spector while listening to Ernst Krenek. Yes, there are differences between the two—even qualitative "distinctions," for those who need to find them. A work like Webern's ten-minute Symphony has little or nothing to do with, say, Van Morrison's ten-minute "Almost Independence Day" but they are both, to this taste, supreme musical experiences and the achievements of one in no way negate the very different achievements of the other. (*Of course* the Webern is wispy and abstracted, but what poetry in those shards of sound! *Of course* "Almost Independence Day" is essentially a fantasia on two simple chords, but what a rich, mysterious cosmos Morrison finds in them!).

I was delighted to be present for the reception that greeted Corigliano last week, and I think *The Ghosts of Versailles* is a distinguished contribution to the repertory. But I hope that musical "accessibility" does not become a fixed credo like "complexity" was a generation ago. In 1970, composers had to defend themselves if they wrote tonal music; today, increasingly, dissonance and complexity are dismissed out of hand. Both responses, if employed automatically, are "knee-jerk" and, it seems to me, morally wrong. Forget the strictures and listen to the music—to *many* musics—and accept it for what it is. The world is full of a number of things and I am glad that the work of Ernst Krenek—thorny, specific, uncompromising, deeply emotive—is among them.

Newsday (1992)

35 *WHO IS SCHUBERT?*

The concerts were not the main event in the 1992 Schubertiade, presented last week at the 92nd Street Y. Instead, our entire understanding of Schubert and his works was called into question—through a challenging series of talks, exhibitions, films and assorted miscellany, curated by Joseph Horowitz. And so this is less a review than a response.

Horowitz commands respect. He is the author of three engrossing books on music—*Conversations with Arrau, Understanding Toscanini* and *The Ivory Trade.* He is the sponsor and mastermind behind the 92nd Street Y's current "Goldfingers" series—recitals by pianists who did *not* take first place in major competitions but are more searching musicians, in many ways, than the ones who did. Moreover, Horowitz, at his best, is one of the few arts writers who can inject social and political theories into his work without making a blithering ass of himself.

Still, there was a certain shrill quality to Horowitz's introductory essay to this Schubertiade—subtitled "Perspectives on Schubert from Weimar to Hollywood"—and, indeed, in the underpinnings of the whole event. Horowitz it would seem, deplores the prettified, happy-go-lucky Bohemian impression of Schubert that has been vouchsafed us through the operetta *Blossom Time*, through popular fiction and through what Virgil Thomson used to call the music appreciation racket. He cites the following passage from an old biography as particularly noxious:

"It is the face of a teacher, but not of a strict one. The hair curls about the brow. No, this is no pedagogue: This is an artist, a true musician, certainly, if not a virtuoso. The gaze is modest: It betokens a man of the people, one who can work and play man to man, one who can casually create beautiful things . . . There is nothing commanding, scheming, nothing pressing, nothing problematical. All flows naturally from heart to heart."

Sure, it's a little much by present-day standards. But what are our own grandchildren going to make of passages such as the following, excerpted from the booklet's description of a symposium called "Schubert the Man—Myth Versus Reality"?:

"Maynard Solomon has recently established that Schubert was part of a subculture of men who engaged in male-male sexual activities. Letters indicate that this aspect of their lives was not simply incidental but that their sense of identity and community revolved around their common sexual orientation."

(Well, gosh. All those linguistic contortions to sidestep a three-letter word!)

Complementing the program was a display of Schubert kitsch in the Y Gallery—old postcards, fanciful drawings of the composer dancing with Josephine Baker, playing guitar, receiving inspiration from on high and so on. One sensed, somehow, an invitation to disapprove of the uses to which Schubert's image had been put. Myself, I loved the exhibit, and think it would make a sweet and amusing book, one that might tell more about the specific appeal of this composer than a dozen maundering academic essays. (And I'd gladly place the Schubert/Josephine Baker print on my wall, right between the Dionne Quints and the "Mr. Peanut" souvenir of the 1939 World's Fair.)

Don't misunderstand me. Great art is profoundly elitist; Indeed, its very essence is selection. Schubert is in "the canon," in "the hierarchy," and considered a Great Composer (to employ Horowitz's—presumably ironic—system of capitalization) for very good reasons. It is no accident—and certainly no conspiracy—that he is generally regarded more highly than, say, Amy Beach or William Grant Still, and to suggest otherwise is mischevious.

But the work of a composer so bounteous as Schubert may be understood on may different levels, not all of them profound, and it seems both naive and wrongheaded to blast the popularizers. I don't think musicians ever took *Blossom Time* very seriously; it was pap for the masses and surprisingly durable and entertaining for what it was. Its existence has in no way augmented or diminished the sustained, solemn pleasures that Schubert's late piano sonatas, last symphony and "Cello" Quintet afford to more discriminating listeners. And while it is certainly interesting to know that Schubert was gay (as a number of the Great Composers were) and that he died of syphillis (as a number of the Great Composers did), I wonder whether our "marginalized" interpretation of his life is any more definitive than the sentimental portrait drawn by the Victorians.

Newsday (1992)

PART III

MISCELLANY

1 PRODIGY: THE ENDURING MYSTERY OF THE MUSICAL WUNDERKIND

I remember coming across one of Robert Ripley's grislier "Believe It or Not" sketches in a book when I was about ten years old. "Boy Dies of Old Age at Seven!!" the headline read, and, in the best Ripley manner, a few terse sentences, augmented by exclamation points, told of a child with a metabolism so quick that he learned to read at the age of two, grew a full beard at four, and withered away and died before his eighth birthday. The boy in the drawing bore a resemblance to the exhausted sage in a thousand New Year's editorial cartoons; spent and wizened, all but shut down, he yet peered at the reader through the eyes of a child.

I probably haven't seen that picture in twenty years, but I've thought about it often and at some length, and I suspect that most prodigies will respond to the image, for they, too, are simultaneously young and old. "Baby's brain and an old man's heart" ran the lyrics to a popular hit of the 70s. Choose your own contradictions—the fingers of a master pianist on a scared boy; an inarticulate shyness hiding the mathematical acuity of a calculator; profound artistic sensitivity that coexists with a remote, disengaged personality. And then, of course, the occasional happy, well-adapted child blessed with a genius that is a delightful plaything.

One afternoon at the Curtis Institute of Music in Philadelphia, nine-year-old Hiroshi Proctor played through Bach's Concerto in D minor on his teacher's piano. Diminutive, cheerful, almost frighteningly self-aware, Hiroshi tilted back and forth upon the piano bench, racing through the music aggressively and skillfully, then branched off into an improvisation of his own that seemed a mixture of cocktail blues, French impressionism and early 19th-century romanticism.

Hiroshi, raised in Japan and Kalamazoo, Mich., began studying the piano four years ago with his mother. She has now moved with him to Philadelphia in order for the boy to attend Curtis. His playing is bright, propulsive, full of energy and intelligence.

Hiroshi is a particularly brilliant prodigy, but he is not the only young artist who has recently aroused curiosity within the musical community. Two years ago, a violinist named Midori, then fourteen years old, created a sensation at Tanglewood by playing Leonard Bernstein's "Serenade" coolly and flawlessly, despite having to swap instruments twice in mid-performance, due to broken strings. The young cellist Matt Haimovitz, a year or two older than Midori, is already a familiar figure on the New York concert stage. And the Juilliard School is keeping a proud, watchful eye on the pianist and composer Wendy Chen, the violinist Gil Shaham and the cellist Melissa Brooks, teenagers all.

There are prodigies in other disciplines as well, although nobody knows exactly how many there are, nor has anyone set up any strict guidelines to distinguish the truly prodigious from the highly talented. But these young geniuses inspire a profound and enduring fascination.

The legends are familiar—young Mozart traveling through 18th-century Europe, playing for crowned heads and commoners; junior chess champions balancing twelve games at the same time and winning them all; mathematical wizards who outpace the pocket calculator. And not only are these prodigies amazingly bright, but they are also, more often than not, genuinely *cute kids,* and therefore automatically exempt from the petty jealousies that extraordinary adults often inspire among their peers.

But nobody knows quite what makes a prodigy. Parental intelligence has something to do with it, of course, as does environment and early training. After that, prodigies are still mysterious, even to themselves.

"I can't tell you what produces an Einstein," Julian Stanley, the director of the Study of Mathematically Precocious Youth, in residence at Johns Hopkins University in Baltimore, said "You can read Einstein's biography until you go blind, but you'll find nothing that explains him, nothing in his background that prefigures his achievement. The prodigies I deal with are so startlingly precocious that their environment had next to nothing to do with their development. We have children who surpass immediately what their parents expected of them, or what anybody thought they were capable of doing.

"There are two kinds of prodigies," Stanley explained, "the ones force-fed from an early age, who are very bright to start out with and then pushed to the limit by their parents in the conscious effort to produce a prodigy. We have a nice term for these mothers and fathers— we call them 'creator' parents. And then you have the extraordinary child who just comes into the world with the desire to make music or with a fascination with square roots."

Stanley says that it is difficult to tell whether the number of prodigies is increasing or decreasing. "We pay more attention than we used to, I think," he said. "We identify them earlier. And they have changed. Prodigies are, one way or another, linked to the zeitgeist. Just now, a lot of children are finding themselves in computer science. This would not have been so easily available to them in previous decades. But they might have done something else that would have been equally impressive."

Prodigies stand apart and they know it. This must be one of the reasons that so many precocious children have a love for the mysterious and the fantastic—science fiction, unicorns, extraterrestrials and the like. They *understand* aliens, for from an early age they are acutely aware of their own difference, their "specialness."

And they don't always like it. "When I was a child, all I really wanted was to be normal," cellist Yo-Yo Ma, now thirty-two, reflected. "From the very beginning, I was set apart because I played the cello. I was concertizing from an early age [he was born in Paris and played his first public recital there at the age of six], and I wasn't allowed to play any contact sports after the fourth grade. So I felt cut off. I always wanted to be liked for myself, not just for my cello playing."

Ma, however, is among the best adjusted of musical prodigies, something he credits to his training at Harvard University, which he attended after seven years in the Juilliard preparatory division. "Attending Harvard instead of a conservatory was the single best thing that ever happened to me," he said. "I started meeting people who were at least as passionate about their fields as I may have been about mine.

"It opened up new worlds," he continued. "It made me get out of myself and think about music as it related to the world, in a very different way than if I had just concentrated on my scales and exercises. It made me much less neurotic. Performances were no longer the only thing in the world. My whole life didn't depend on the success or failure of a performance, and I could see that there were other things that I could be if I were not a musician."

Not all prodigies realize that they have such an option but, to their credit, many conservatories are trying hard to be more receptive to the uneven development of their students. Educational programs for the gifted have been instituted in many schools across the country. "Some prodigies are not directed, and schools and parents can help provide that direction," Stanley said. "Most of the prodigies I know have high IQ scores, but there are two things to keep in mind there: First, that you can't major in IQ, and, second, that for fifty cents and a high IQ you can get a fifty-cent cup of coffee."

Gary Graffman, the artistic director of the Curtis Institute, understands the difficulties of a prodigy better than most: He entered

Curtis as a student at the age of seven, the youngest pupil in the school's sixty-year history. A celebrated pianist for many years, who was forced to retire because of a physical problem with his right hand, Graffman has written an engaging autobiography, the title of which could serve as a mantra for many musical prodigies: *I Really Should Be Practicing*.

"Unlike some other conservatories, Curtis doesn't have a preparatory department," Graffman said. "If you're good enough to get into Curtis, you get into Curtis, no matter how old you are." He says that most of the students at Curtis are "normal, everyday kids who just happen to be very talented. You must realize that almost every pianist, every violinist, every musician is a child prodigy in one way or another," he added. "They all start off between the ages of three and five, maybe six or seven at the outside. It really isn't going to happen for them otherwise. If you are going to flourish during your teens, you have to have a pretty good technique by then. Because technique is just the beginning."

Indeed it is. But sustained individual creation is rare among youngsters. One finds a particular concentration of prodigies in the fields of music, mathematics and chess. It has been suggested that these disciplines—built upon complex rule structures rather than dependent upon extensive personal experience—are perfectly suited to precocious children, who learn the rules and regulations, then work with them at a high level of achievement. Most of the famous prodigies are *re-creative* artists; in some cases, they are little more than brilliant mimics, able to do anything that is asked of them, but nothing more, nothing original.

For while it is true that Mozart was writing professional-quality music by the time he was eight or nine (most scholars consider his earlier compositions to be at least partially the work of his ambitious father, Leopold) none of them were particularly memorable or distinctive, and they would likely have been forgotten had it not been for the composer's later achievements. Mozart's first real masterpieces date from his teen years, as do those of another musical *Wunderkind*, Felix Mendelssohn (who produced two of his most beloved works, the overture to "A Midsummer Night's Dream" and the string Octet, in his seventeenth year).

The same generalization applies across the board. No great literature has been produced by a prodigy (the poems of Chatterton and the young Keats are the work of teenagers), and when we speak of "boy genius" filmmakers, we refer to relative oldsters such as Orson Welles, Bernardo Bertolucci and Peter Bogdanovich, all of whom made their reputations while in their early twenties.

Julian Stanley observes that children have made no important mathematical discoveries: "The earliest major achievement that I know of was the theory of 'least squares' developed by Carl Friedrich Gauss when he was about eighteen. Before that, he had done some clever things—correcting his father's arithmetic and so forth—but nothing vital."

Not all of the performing prodigies continue to develop: The late recordings of Jascha Heifetz are not markedly different from those he made at sixteen in 1917, the year of his Carnegie Hall debut. Many believe that Yehudi Menuhin, who was world famous by the age of twelve, never quite recovered from his imperfect training, that his early playing was inspired by a youthful poise he was never able to fully recapture in maturity.

Some prodigies, pushed too far too fast, ultimately disintegrate. There have been few violinists so naturally gifted as Michael Rabin, who made a brilliant name for himself as a teenager, then lost control during his twenties, and died in an accident at the age of thirty-five, just as he was recovering his mastery—indeed, earning it consciously for the first time. For other musicians—Ma, Itzhak Perlman and the late Glenn Gould among them—the growth process is less painful.

All agree that the attitude of parents is tremendously important to the development of youthful talent. "Some parents push their children very hard; other parents let them grow naturally," Graffman said. "My parents used to insist that I get away from the piano now and then. They'd make me go out and play in the park, into the street to play stickball. But we do see a lot of aggressive parents trying to attract attention to their little geniuses. They used to be called Jewish mothers, but now, in the conservatories at least, they're usually Oriental mothers."

Wendy Chen, now seventeen, who has been studying piano in Juilliard's preparatory division since she was seven, and has matriculated into the college for this fall, learned about music from her mother, and her initial association is charged with deep emotion. "My sister was born when I was four and a half," she recalled, "and after that, for a while, the only time I really got to spend with my mother was when she was teaching me the piano. I really cherished those moments and they taught me to love music."

Wendy is a prolific composer as well as a pianist; she began to write her music down when she was seven. She recently played her piano concerto, an ambitious, fully orchestrated work, at the National Theater in Taiwan. She has also written choral and orchestral pieces, many compositions for solo piano, a musical homage to Charles M. Schulz, the cartoonist, and two operas, *Hello Diary* and *Fighting Problems*, the latter of which, she promises, "is not as violent as it sounds."

Her taste in music is catholic; she names Chopin, Mozart, Liszt, Stravinsky and Andrew Lloyd Webber as her favorite composers, and one can hear their influence on her music, although she is rapidly developing a voice that is her own. Wendy is refreshingly candid about the reasons for her interest in performing: "I love the attention, and I love to make other people feel good." Her melodies are infectious, her sense of harmony conservative but not unsophisticated.

Wendy had a happy childhood—"There were lots of friends, and my family was very supportive. I don't think I was overprotected. I played a lot of volleyball. I shudder a little about that now, because it's just about the worst sport for your hands, but I was on a team and it was great."

No problems here; although nobody can predict who will be a star of the first order—too much depends on circumstance—Wendy seems cheerful, fit and eager to meet the future.

Graffman believes that Hiroshi, who has already begun to concertize—he recently played a concerto in Kalamazoo—is also particularly well adjusted. "His mother is very level-headed," he said. "She doesn't push him. She lets him discover the world on his own terms."

It's a fine line. Ma, now a father of two, is determined to be careful in the way he raises his children. "I hear parents telling their kids that they, too, can be famous soloists if they work hard enough," he said. "That, to me, is the worst thing you can do to a child. If you lead them toward music, teach them that it is beautiful, and help them learn, say, 'Oh, you love music, well, let's work on this piece together and I'll show you something,' then that's very different. That's a *creative* nurturing. But if you just push them to be stars, and tell them they'll become rich and famous—or, worse, if you try to live through them—that is damaging. I was lucky. For all of the pushing I received—and it was considerable—my parents had a high regard for learning and that saved me."

Like several other ex-prodigies, Ma recommends a book called *The Drama of the Gifted Child* by Alice Miller. "She is a psychologist, and many of her patients are people who achieved a great deal in their early years. And they come to her in midlife and tell her that everything is going terrifically well on the surface, but that they feel empty inside. And, often, it turns out that these patients have subjugated their own identities and remade themselves in an effort to please their parents. You find this sort of struggle a lot with gifted children."

One example, Ma said, was composer Igor Stravinsky. "He had an older brother whom his parents apparently preferred, and Stravinsky determined that when he grew up to be a man, he would prove his worth. But it was a hollow victory for him, because by the time he was world famous, both of his parents were dead.

"There are two sides to it, of course," Ma said. "Some children have a fire in their belly, a determination to get things done, that is individual and has little to do with parents—or anybody else, for that matter. Some people soar toward a specific goal. The prodigies who grow and prosper are those with an instinct for survival—the determination to get out of any unhealthy situation and try and find answers."

Julian Stanley said, "It is important to remember that gifted children are still children. They should interact with their classmates. They should have an athletic activity. And they should be exposed to all kinds of intellectual stimulation."

Hiroshi seems to have thought a good deal about all sorts of issues. And he loves to talk. Interviews, it is clear, are tremendous fun.

"Benjamin Franklin lived in Philadelphia," he said. "Did you know that he invented the rocking chair? And the take-out library? That's the thing I like best about Philadelphia. The library across Rittenhouse Square from Curtis. I go there all the time. I learn things and I work on my essay, 'The Exploration of Space.' I'm up to the 42nd line. I guess that isn't too far, so I've got a way to go."

He says that he has no religious beliefs, is "undecided" about politics and that his heroes are Isaac Asimov and Leonard Bernstein. "Bernstein brings such strength to everything he conducts, and yet he can be so gentle, too. He's very famous, you know."

He plans to divide his energies between science and music, and has already evolved a revisionist theory about the "big bang," which was rather too arcane for my unscientific mind to follow, but on which he elaborated with vivid hand gestures and boundless self-confidence for several minutes.

"I will have a good life," he concluded. "I will play some concerts, and invent things when I am not playing."

The Newsday Magazine (1988)

EARLY MUSIC, LATELY

In the late 1950s, Sir Thomas Beecham, the wealthy and unlikely British conductor, decided to record Handel's *Messiah.* He began by omitting all of the arias and choruses that, in his opinion, did not measure up to the composer's best. Then he spiced up Handel's chaste Baroque instrumentation with a battery of tubas, trombones, cymbals, xylophones and other sore thumbs, engaged a gigantic chorus and equally sumptuous orchestra, and produced an exuberant, idiosyncratic (and loud) *Messiah*—a joyful noise, indeed!

Even as far back as 1959, when the record was released, some listeners were dismayed by Beecham's cavalier approach and, over the course of the past three decades, the purists have prevailed. Recently, the most honored—and commercially successful—recording of *Messiah* is Christopher Hogwood's performance with the Academy of Ancient Music, a very different affair.

In lieu of Beecham's huddled masses, Hogwood utilized a choir of thirty sopranos and altos—all boys—and a small orchestra, *sans* tubas. Beecham's pomp and grandeur was replaced by an exquisite transparency. Hogwood made no arbitrary cuts; indeed, he examined all of Handel's various editions of the score, then elected to perform the "Foundling Hospital Version of 1754," the fine points of which were elucidated in a lengthy, scholarly program note. The result was a fleet, airy *Messiah* that sounded utterly different from the recordings and performances that most of us grew up on.

What had happened in the meantime? Well, "Authenticity" had happened—a flowering of musicologically informed interpretations of Baroque and Classical era works that proved appealing to musicians, critics, audiences, promoters and record companies. Indeed, when the history of performance in the 20th century is written, this will likely be remembered as the era of authenticity.

This was the era that brought the musical scholar out of the museums and libraries and into Lincoln Center. It revived gut strings, valveless horns, "natural" pitch and small, focused specialist ensembles. It made superstars out of such artists as Trevor Pinnock, John Eliot

Gardiner, Roger Norrington and Hogwood. It brought obscure Handel oratorios, half-forgotten Rameau operas and concertos and cantatas by the "other" Bachs to public attention.

And more. Archiv Records, which began more than three decades ago as a small semischolarly offshoot of Deutsche Grammophon, suddenly became a central (and highly profitable) division of its parent company and, indeed, one of the hottest labels in the industry. L'Oiseau Lyre, for which Hogwood records, experienced a similar surge of popularity, after years on the periphery of the budget charts. Ultimately, the early music revival changed the music business—and a lot of minds, as well.

Recordings reach a vastly larger audience than concerts, so I have discussed them first. But there is no shortage of "authentick" early music in the concert hall. In August, the British conductor Roger Norrington brought the "Beethoven Experience" to Pepsico Summerfare in Purchase, New York, one hour north of Manhattan. Beethoven's life and works were examined through papers, presentations, rehearsals, workshops and performances. "The Beethoven Experience" was adapted from a wildly successful series of such "experiences" that Norrington has presented on the South Bank of London—weekends devoted to Haydn, Mozart, Beethoven and Berlioz that have played to delighted overflow crowds.

London remains the center of the early music revival; there, Norrington, Hogwood, Pinnock and Gardiner pursue their research and mount what Norrington prefers to call "historically informed" (as opposed to the presumptuous "authentic") performances of Baroque and Classical music. But these artists visit the United States regularly and rarely play to an empty seat. And Pinnock has now made a commitment to working in America, by accepting the directorship of a new orchestra, the Classical Band.

At their best, performances of Baroque- and Classical-era music played on original instruments provide a fair approximation of what composers and audiences might have heard in the courts and palaces of 18th-century Europe. The ensemble will be smaller than a typical philharmonic; orchestral textures will be wispy and delicate; the violins, violas, cellos and basses will be strung with gut and played with a minimum of vibrato. The woodwinds will probably be made of real wood; the horns will not have valves. The tempos are likely to be fast, the rhythms clipped, the interpretations determinedly anti-Romantic, with none of the swells and exaggerations that we have come to associate with that era.

The best of these performances are little short of revelatory. I recall a 1982 recording of Vivaldi's *Four Seasons*, featuring Pinnock and the English Concert, with Simon Standage on violin. Here they were, the most hackneyed concertos in the Baroque repertory, dashed off by every band and fiddleplayer in the business, recorded on the Moog synthesizer—even adapted as a film score by Alan Alda. And suddenly, in the hands of Pinnock and Standage, the music seemed brand-new, crisp and clean, the distortions of past performances scrubbed away, the sound tapered, sinuous and sweet.

Unfortunately, not all of the manifestations are so engaging. Too many early music groups have played with a grim sense of messianic duty, their ornamentation self-conscious and excruciatingly proper, their instrumental timbres dry as dust. One finds oneself wishing that the players would cut loose now and then, and let their instruments sing out, rather than controlling them so carefully, for this can easily degenerate into a music of inhibition. I have heard impeccably authentic performances that seemed learned rather than felt, and came across as little more than a succession of enervated, undernourished sounds— even when the composer, whatever his original forces may have been, clearly wanted the skies to open.

For example, Otto Klemperer made a famous recording of Bach's *St. Matthew Passion* about twenty-five years ago. From a musicological point of view, we now know his reading to be problematical—it is massive, heavy and semi-operatic, realized with a huge orchestra and chorus. But the terror, the pity, the grave power of this drama come across with an unforgettable intensity. Compared to this monolith, most of the recordings for smaller, more stylistically appropriate ensembles seem cautious and scholastic, almost trivial.

The revival of interest in such instruments as the harpsichord and fortepiano is central to the movement toward stylistic authenticity. The debate about whether Bach's music should or should not be played on the piano will probably stretch on for years, and this listener, for one, is sick to death of it. While I'm delighted that we have so many fluent harpsichordists before us today, who would want to be without Glenn Gould's charged, highly original piano renditions of the keyboard works, or the more recent performances of Andras Schiff?

Of all the great composers, Bach was the least concerned with the actual *sound* of his music; indeed, he was one of the premier self-transcribers of all time. He happily and successfully rearranged his violin concertos for the keyboard and recycled choruses into *sinfonias*. His principal interests seem to have been melody and structure, rather than the sound of his instruments; for this reason, his music survives a multiplicity of processing. Bach's compositions work on the synthesizer, in versions for jazz chorus, even in arrangements for saxophone quartet. And who can say whether the composer would or would not have approved?

In recent days, the fortepiano, a precursor of the piano, has returned to the concert halls, and not always to good effect. "Someone left a cake out in the rain" ran the lyrics to a hit song of the 1960s and, at too many early music concerts, it sounds like somebody then tossed an old piano into the storm to comfort the sodden angelfood. But the best fortepianists add something crucial to our understanding of early keyboard music. Steven Lubin, for example, with his Mozartean Players, makes *music* with the instrument, capitalizing on the gray, somewhat closeted sound and uneven registration, generally transcending what can seem, in other hands, insurmountable limitations. His playing is clean, unmannered and deeply musical, virtuosic but never ostentatious.

Ultimately, the movement toward playing Baroque- and Classical-era music on so-called original instruments is more about a state of mind than anything else. Most violin soloists play 18th- and early 19th-century instruments, after all, whether they are performing Bach or Berg. Likewise, many, even most, of the harpsichords and fortepianos that we regularly encounter in concert are actually modern duplications of early instruments. It is the way these instruments are *played* that makes all the difference.

This is one of the reasons that Norrington has attracted so much attention. He brings good, old-fashioned "blood'n'guts" to a style of performance that sometimes seems pedantic and desiccated. And his rendition of the Beethoven Eighth Symphony, which he played at Carnegie Hall last April (and has recorded for EMI Angel as part of an ongoing survey of the complete symphonies), was a triumph.

Norrington's Beethoven is swift, streamlined, objective, elemental. His conception is based on dance and folk song, rather than metaphysical rumination; it is of the Earth rather than the heavens. One is always aware of the music's pulse, and Norrington phrases in broad strokes, without italicizing. I cannot recall a more exciting performance. It seemed, in the words of a famous advertising campaign, "shot from guns"—one joyful trajectory from start to finish.

Norrington has a simple explanation ready for the differences between his conducting and that of the other early music specialists. "I don't play the harpsichord," he said. "I don't come from that precious, self-conscious, background. Most of the early-music people are lutenists or harpsichordists. I was a violinist and a singer.

"More to the point, I am a real conductor," he continued. "God, that sounds pompous, but I have a point. I didn't come to this directly from a library or a chamber group. I've been conducting orchestras for 25 years." (Norrington's season also includes performances of Britten's *Peter Grimes*, Verdi's *La Traviata*—even *H.M.S. Pinafore*.)

There is little doubt that Norrington is currently the popular and critical favorite among the early music players, and that he is indeed a "real"—and often very exciting—conductor. But the way in which many of the early champions of Hogwood, Pinnock and company are now using Norrington as a club to beat their favorites of yesteryear is emblematic of an unpleasant side to this business.

Nothing outrages small, passionate cult audiences more than the sudden acceptance of their heroes into the mainstream. Followers can forgive almost anything else—mannerisms, stylistic reversals, failed experiments—so long as the password remains a secret and that exotic clandestine snob appeal continues to tingle.

For example, it has lately become fashionable to denigrate Christopher Hogwood and the Academy of Ancient Music. The group has become too popular, made too many records (including sets of the complete Mozart symphonies), been played too many times on FM radio—in short, has established itself as Everyman's early music group. And so many of those musicians and critics who once hailed the Academy as a forerunner of a "brave old world" have scurried off, in search of other idols.

It is important to give Hogwood his due and to recognize the fact that much of the reaction against his work is just that—reaction. There was a similar reversal of critical opinion about the late Karl Richter, who taught an entire generation of people how to think about the Baroque with his streamlined Archiv recordings of Bach and Handel in the late 1950s and early 60s. Richter, who displaced the style of grand, romanticized Baroque performances of such men as Stokowski and Klemperer, with performances of a near-Verdian propulsion, was himself displaced by Pinnock, Gardiner and Hogwood.

One of the tests of a masterwork is its ability to withstand a multiplicity of interpretations. I have enjoyed and learned from the work of Norrington, Pinnock and Hogwood, and I expect to continue to do so. But I have also learned from Beecham, Stokowski and Klemperer. Although it remains unfashionable to say so, their oversized Baroque and Classical performances sometimes came closer to the essential spirit of this music than many a latter-day musicological realization.

The tide may be turning, however. Last year, Andrew Davis recorded a *Messiah* for EMI Angel, with a gargantuan chorus and orchestra, backed by a big, booming organ and fronted by one of those lush, plummy, maternal altos who seem ready, willing and able to succor the world. Purists shuddered but Davis was in fact resuscitating a time-honored tradition. After all, why shouldn't there be a good, old-fashioned recording of *Messiah,* stuffed in the Victorian manner, amidst what is now a plethora of authentic Baroque performances?

Sir Thomas would have loved it.

Elle (1989)

3 THE MOZART BICENTENNIAL

On December 5 it will be exactly two centuries since Wolfgang Amadeus Mozart died in Vienna and so, in musical circles, 1991 has been designated "the Mozart year."

Over the next few months Mozart, already one of the most familiar composers in the literature, will become downright ubiquitous. On what would have been the composer's 235th birthday, Lincoln Center will begin a 19-month observance of the bicentennial, with performances of everything Mozart is known to have written. It will include operas (the Met will present the seven in its repertory), dance performances, symphonic concerts, chamber music, solo recitals, lectures, symposiums and broadcasts—some 500 events in all.

Meanwhile, the Netherlands-based Philips label is in the midst of recording an integral edition of all of Mozart's works, while single commemorative albums from other companies arrive daily. WNET has just presented Peter Sellars' idiosyncratic stagings of the three operas Mozart composed to texts by Lorenzo Da Ponte—*Le Nozze di Figaro, Don Giovanni* and *Cosi fan tutte.* More telecasts will undoubtedly follow, while the publishing houses will offer a mix of coffee-table volumes, popular biographies and the occasional serious contribution to Mozart scholarship.

And, of course, the journalists will weigh in. If past performance is any indication, we can soon expect dewy-eyed, decline-of-the-West ruminations on sublimity and early death from Sunday sections around the country (one famous example actually used Mozart as a club with which to beat Elliott Carter). The record magazines will have a field day, with lengthy appreciations and "which-*Cosi*-should-you-buy?" articles. There will be features in the major weekly news magazines— the magazines that rarely write about concert music anymore, trying to say everything at once and ending up not saying very much at all: This-that-Vienna-*Wunderkind*-"Amadeus"-and-whatever-one-thinks-of-Mozart-he-has-given-us-something-to-think-about. Everyone will find an angle: I wouldn't rule out stories in *Boxing World, Popular Mechanics* and *National Geographic.*

By now, the reader will have detected a certain cynicism seeping through this newsprint. Rest assured that it is in no way directed toward Mozart's music. Indeed, if I were told I had only an hour to live, I think I'd ask for a recording of the finale of *Nozze di Figaro* to sing me away; this serene, Olympian exaltation of forgiveness has long seemed to me an expression of the highest human values.

And (to employ an equally unlikely but rather less bloody-minded hypothetical situation) if I were banished to the moon tomorrow and could take only a handful of recordings, I know that my selection would include performances of the last six symphonies (along with one of No. 29 in A); the string quartets dedicated to Haydn; the Divertimento for string trio (K. 563); an assortment of chamber music; the "Great" Mass in C; the "Ave Verum Corpus"; the last ten or so piano concertos; the last three violin concertos; the Sinfonia Concertante for Violin and Viola; and several of the operas. Back on Planet Earth, there are few more satisfying tactile and emotional challenges than speeding through a Mozart keyboard sonata.

But the Mozart Bicentennial—like the Bach-Handel-Telemann triple tricentennial in 1985 and the centennial of Tchaikovsky's death that will unquestionably come our way in 1993, and all the other predictable anniversaries into the next millennium—leaves me with mixed feelings. Never mind that we hear a good sampling of Mozart's masterpieces at Lincoln Center every summer (the "Mostly Mozart" festival is aptly named). Never mind that there are dozens of satisfying recordings of the composer's best music. Never mind that the two most significant figures in American musical history—Leonard Bernstein and Aaron Copland—have both died within the last three months, and nobody has proposed any kind of "compleat" survey of their work.

No, we are encouraged to genuflect to Mozart once again while our arts organizations use the composer as an excuse to further limit their already timid programming. Imagine—pure fantasy, now—what might have happened if the powers at Lincoln Center had decreed instead that 1991 would be "the premiere year" and requested that all visiting ensembles play a new or unfamiliar work. There would still have been plenty of time left for Mozart, and the art of music might have advanced a step or two.

Moreover, I suspect that new compact discs of the complete Mozart up on the shelf may not be much more than handsome furniture for most listeners—the modern equivalent to those matched, unread editions of Bulwer-Lytton, Carlyle, Longfellow and Balzac so beloved of our grandparents. Is it really necessary to have every fledgling, pre-teen symphony, every wind divertimento that Mozart composed at hand?

And wouldn't the mysterious young pianists who have rushed in to give the world their recorded thoughts on the sonatas have been better advised to investigate the many corners of the repertory still unexplored, where, unchallenged, they might make a better showing?

So two cheers for "the Mozart year." In the next months, we will be surrounded by a great deal of beautiful music, played by honored artists from around the world. Still, Mozart really needs no such platform. We know him; we love him; he is part of our experience; he will, one hopes, be with us forever. But music is a living art and, right now, there are distinguished composers working in the United States who have yet to have a single piece sponsored by Lincoln Center. Only through performance can we separate the wheat from the chaff, the distinctive from the purely academic; only through performance can we begin to discern the masterpieces of tomorrow. When will our major arts organizations do their duty to their own time, so that our descendants may someday celebrate the centennials of some late twentieth-century composers?

Newsday (1991)

4 THE NEW AGE OF BRIAN ENO

In his novel *Vineland,* Thomas Pynchon speaks dispargingly of "New Age mindbarf dribbling out of the P.A. system." That pretty well sums up the way most musicians—and critics—feel about the genre.

But what is "New Age" music? Depending upon the person you ask, it can encompass everything from Vivaldi to Steve Reich. Searching through Tower Records for some discs by Brian Eno the other day, I discovered that his entire oeuvre was now catalogued under "New Age." The placement is certainly improper for his early rock albums—"Here Come the Warm Jets" (1973), "Taking Tiger Mountain by Strategy" (1974), "Another Green World" (1975) and "Before and After Science" (1977)—which strike me as among the most engaging and original collections of popular songs from that era, spiky and eccentric rather than cosmically soothing. But does the phrase apply to his later work?

One friend thinks so. "New Age with better program notes," he sniffed. There is no doubt that Eno's later albums have much in common with the New Age aesthetic: They tend to be constructed around drones and sonic colors; fashioned with a deliberately limited number of musical "events"; put together with a minimum of visceral aggression (indeed, generally avoiding any dramatic tension except a sense of mystery and expectation) and washed with an electronic sheen.

Yet Caroline Records' compact disc reissue of the Eno catalogue is worth a listener's attention. And, while I regret the fact that Eno has seemingly abandoned more-or-less traditional popular music (the wonderful arching melodies on "Warm Jets"! the wry, bizarre orchestration on "Tiger Mountain"!) the fact remains that the best of his later music contains rewards of its own. I'll leave it up to the listener to decide if these discs are "New Age" or not but they most certainly deserve more than a blithe, categorical dismissal.

The earliest signs that Eno was moving away from rock came with two albums he made with the guitarist Robert Fripp—"No Pussyfooting" (1973) and "Evening Star" (1976). The first of these is the better of the two: Fripp lays down slithering and majestic guitar melodies over Eno's electronic insect hum. "Discreet Music," a solo venture from

1975, contains three haunting variations on the Pachelbel Canon (another seminal work in the New Age history books), distilled to a liquid, pulsing essence.

During the late 1970s and early 1980s, while Eno was busy producing recordings for such groups as Devo and the Talking Heads, he continued his own non-rock collaborations with lesser-known musicians. One, an album called "Possible Musics" (1980) with the avant-garde trumpeter Jon Hassell, is worthy of investigation for the manner in which it combines African stylings, electronics and a sort of personalized minimalism. Another, "Day of Radiance" (1980), featuring the mantric, primitive dulcimer playing of Laraaji (whom I remember listening to in Washington Square Park one late summer evening a dozen years ago as he pounded blissfully on his homemade instrument) alternates wafting meditations with three other compositions (known, collectively, as "The Dance") which might be described as the perfect New Age exercise music—curious, obsessive and vigorous.

I find Eno's two records with Harold Budd, "The Plateaux of Mirrors" (1980) and "The Pearl" (1984) the least interesting in the series: Here, a perfumed, cosmetic prettiness prevails. Much better is a disc created with Daniel Lanois and Roger Eno, "Apollo: Atmospheres and Soundtracks" (1983) in which Eno's electronic landscapes are enlivened with a strange (but satisfying) infusion of Country/Western.

Best of all, perhaps, is "Thursday Afternoon" (1984), a solo album that contains only a single, hour-long composition. Virtually motionless but constantly engrossing, "Thursday Afternoon" began as the soundtrack to a video that Eno created for Sony Japan. "It is an even-textured, spacious and contemplative piece in which several musical events appear and recur more or less regularly," C.S.J. Bofop observed in his unusually perceptive liner notes. "Each event, however, recurs with a different cyclic frequency and thus the whole piece becomes an unfolding display of unique sonic clusters." Heard in the proper frame of mind, "Thursday Afternoon" inspires, paradoxically, both a relaxation and a quickening of the senses.

"New Age with better program notes"? Maybe. But, if so, I must count myself a belated, discriminating but genuine convert.

Newsday (1990)

5 BRIAN WILSON

In the summer of 1984, impelled by a passion for landmarks and caught up in renewed admiration for the Beach Boys' "Pet Sounds" album, a cassette tape of which had provided the soundtrack for our drive down the Pacific Coast Highway, my wife and I made a pilgrimage to Hawthorne, California, the Los Angeles suburb where Brian Wilson grew up.

There is no surf in Hawthorne. Slapped together all but overnight at the end of World War II, this small city is several miles inland, a settlement of practically indistinguishable tract houses, supermarkets, donut shops and filling stations, set amidst the palm trees and the smog. Although the drive to the beach is not an arduous one, Hawthorne is much closer in spirit to South Central Los Angeles, an area now notorious for Crips, Bloods and drive-by shootings.

The timing for our visit was fortunate. 3701 West 119th Street, at the corner of Kornblum Avenue, was already slated for demolition, along with the surrounding blocks. This entire section of northern Hawthorne would soon be cleared—as expeditiously as it had been assembled—to make room for the Century Freeway, another congested artery intended to stave off for a few more years the complete immobilization of Los Angeles. We had imagined that a plaque might adorn the Wilson house. On the contrary, it was boarded up, the grass was dead and garbage lay in the breezeway. The only sign of life was a big black dog, a patchy aggregate of teeth and ribs, who snarled at our car.

This was Hawthorne, from which the young Brian Wilson emerged, almost thirty years ago, with a musical style derived from Chuck Berry and the Four Freshmen, and the seductive vision of an idealized California—girls on the beach, a pristine coastline, open highways, the warmth of the sun...

Myself, I never bought the California myth (a poor swimmer who never learned to drive, I was pretty much excommunicated from the start). Moreover, there is little to say about the Beach Boys' later years (lame albums, formulaic public appearances). Indeed, were it not for the monolithic presence, seen or unseen, of Brian Wilson (and he is not now performing with the band), the Beach Boys would likely have been relegated to Holiday Inn "flashback nights" years ago.

Still, the group's best recordings—those from the mid-1960s through the early 1970s, composed, arranged and produced by Wilson—are enduringly satisfying. Capitol Records has begun to reissue all the early Beach Boys albums, the last large oeuvre of unquestionably important rock LPs to make it to compact disc. Included in the initial release are their primitive first efforts "Surfin' Safari" (1962) and "Surfin' USA" (1963)—the title song of the latter a shameless appropriation of Chuck Berry's "Sweet Little Sixteen"; the transitional, slightly more sophisticated "Surfer Girl" (1963) and "Shut Down, Volume Two" (1964)—the first volume of "Shut Down," issued in 1963 and well forgotten, was a compilation disc featuring the Beach Boys and several other artists in car songs; and the baroque, full-blown glories of "Pet Sounds" (1966), a nearly perfect record, reissued with two additional cuts.

At the same time, on a bootleg compact disc from Japan, comes "Smile," the most famous rock album never made, a collection of songs and miscellaneous fragments on which Wilson was working before the breakdown he suffered in 1966 or 1967 and from which—public relations to the contrary—I think he has yet to recover.

"Smile," the lost record, promised time and again but never released, has been the subject of books, lengthy articles and much fantasy. Stray tapes from the sessions have circulated among collectors for years, but the Japanese disc contains a full hour of previously unreleased material and provides probably the closest approximation of Wilson's original conception that we will ever have. With its strange, elliptical little canons—quirky miniatures that explode in the mind—and the distinctive harmonic sense that Wilson had begun to evolve, as well as his exquisite sensitivity, unparalleled among rock musicians, to the pure *sound* of a record, this reconstituted "Smile" reveals Wilson, once and for all, as one of America's great natural talents.

That talent is apparent, however fleetingly, even on the first albums. The beat is rudimentary and the harmonies augmented barber-shop, but there are flashes of an intensive lyricism that became more pronounced on later recordings. "Surfin' Safari," "Surfin' USA" and "Surfer Girl" made the Beach Boys famous—indeed, made them the most popular rock group in the world before the advent of the Beatles—and, while they now seem crude to most listeners, they contain a disproportionate amount of the band's best-remembered hits.

"Shut Down, Volume Two" marks an advance. While there is a lot of filler on this album—including some dismal locker-room humor ("Cassius Love versus Sonny Wilson"); songs constructed to fill what were by now familiar formulas ("This Car of Mine," "In the Parkin' Lot"); and cover versions of "Why Do Fools Fall in Love" and "Louie Louie"—it also contains "Don't Worry Baby" and "The Warmth of the Sun," expansively lyrical, sumptuously arranged, meticulously produced songs that have an emotional resonance far beyond anything Wilson had written so far.

"Pet Sounds," that ever-fresh, singularly affecting study of a youthful love affair, charted from idealized beginnings through disillusionment and ultimate dissolution, is often hailed as the first "concept" album. In essence, it's a unified song cycle, broken up only by the interpolation of "Sloop John B," the group's recent Top–40 hit, taped several months before the rest of the record.

On "Pet Sounds," Wilson consolidates and expands the sonic world he had begun to explore with three albums of new material, each one better than the one before, that the Beach Boys had released between August 1964 and June 1965. "All Summer Long," the first of these, marks the fullest expression of the group's early style—the California myth at high tide—and alternates between spirited rockers ("I Get Around," "Little Honda") and increasingly introverted ballads. "The Beach Boys Today," which followed, is the most personal album Wilson had yet made and many of its songs betray a depressive vulnerability ("Please Let Me Wonder," "When I Grow Up," "She Knows Me Too Well"—even the uptempo "Help Me Ronda" is a plea for rescue).

"Summer Days (and Summer Nights!!)" is a sort of synthesis of the two previous releases, fashioned with unprecedented care and near-symphonic production values. While the earlier records were fairly straightforward documents of the way five young men sang their songs (to accompaniments of varying sophistication), on "Summer Days," for the first time, Wilson's fascination with sound comes to the fore. The album is distinguished by a startling textural richness—vast, overdubbed, celestial choruses; carnival organs; tinkling bells; lunatic screams; deliberately strained, out of tune voices; castanets buzzing like so many cicadas—that never distracts from the vitality of Wilson's songs but, rather, enhances their power.

On "Pet Sounds," one finds Wilson's attention to timbral detail mated to unfailingly imaginative, painfully tender melodies. The harmonies are often equally surprising; who could have predicted the complicated, detached bass line in "I Just Wasn't Made for These Times," the sudden key changes in "That's Not Me," or the impressionistic meandering of "Let's Go Away for Awhile"? ("Here's a good way to describe it," Wilson said of the last composition. "Try to hum it.")

After "Pet Sounds," something happened to Brian Wilson; exactly what is still a matter of conjecture. He isolated himself from all but a few intimates and worked steadily on what he believed would be the Beach Boys' *magnum opus,* a record called "Smile." The two most common explanations for the fact that it never appeared are dissension with other members of the band (who were said to be disheartened by the relatively poor sales of "Pet Sounds" and wanted no further experimentation); and the use of psychedelic drugs, which, by all accounts, frightened Wilson badly.

"Smile" was nothing if not ambitious. A total of eighteen songs were recorded, all or in part, over the course of seventy-two studio sessions (an incredible extravagance, by the way; complete symphonies are regularly recorded in a single sitting). Capitol, primed for a hit, printed more than 400,000 album jackets. But the release date was delayed, and delayed again.

Eventually, something called "Smiley Smile," with a different cover and different lineup of songs, limped into the stores. It is probably the most disappointing album of Wilson's career. Although "Smiley Smile" contains two magnificent singles—"Good Vibrations" and "Heroes and Villains"—overall, the record is diffuse and silly, a giggly exercise in featherweight humor, barely hinting at the fractured majesty of the "Smile" sessions.

The Japanese bootleg, on the other hand, contains fifteen minutes of an elongated "Good Vibrations"; episodic snatches from the outtakes of "Heroes and Villains"; serene fragments called "Holidays" and "Barnyard"; a melancholy version of "You Are My Sunshine," retitled "Old Master Painter"; and subtler arrangements of the songs "Wind Chimes" and "Vega-Tables" than the ones that appeared on "Smiley Smile." "Smile" was to have contained a suite called "The Elements," broken into movements depicting earth, air, fire and water. "Mrs. O'Leary's Cow," the "fire" section, is included here. This strange, two-minute cut is said to be one of the main reasons why "Smile" was never released. Wilson apparently decided that the song was causing fires around Los Angeles simply by existing and he is said to have tried unsuccessfully to burn the tapes.

Several of the key songs from "Smile" were eventually released on other records ("Our Prayer" and "Cabinessence" on "20/20"; "Cool Cool Water" on "Sunflower" and "Surf's Up" on the album of the same name). From them, from the Japanese bootleg, and from "Good Vibrations" and "Heroes and Villains," we can get a pretty good idea of what "Smile" might have been. Wilson seems to have been moving toward a remarkably abstracted ideal of song construction; toward gatherings of minute, luscious, elaborately produced modules of sound that, somehow, fit together to form a whole. Twenty years on, the best of these recordings—"Cabinessence," with its vivid vocal depiction of frozen winds; the lyrical and melodic densities of "Surf's Up"; the vaporous, ethereal "Cool Cool Water"—have lost none of their power.

This, however, marked the end of Wilson's experiments. "Wild Honey," which followed "Smiley Smile," is a wonderful record, lean and joyous, but it is a retreat into rhythm and blues and simple, straightforward pop. After "Wild Honey," the other Beach Boys started writing songs; such albums as "Friends" (1968), "20/20" (1969), "Sunflower" (1970), "Surf's Up" (1971), "Carl and the Passions" (1972), "Holland" (1973) and "15 Big Ones" (1976) are very much team efforts. Carl Wilson, over time, proved himself a distinctive and often inspired composer, but none of these discs have the consistency of the group's best early records.

Still, Brian Wilson provided at least one or two beautiful songs for all of them (I except "15 Big Ones," the group's "comeback" recording, which strikes me, now and forever, as completely without merit). "All I Wanna Do," on "Sunflower," is a swirling fantasia for rock band and echo chamber, accomplished so skillfully that it transcends what might have seemed gimmicky origins. "Till I Die," on "Surf's Up," is a secular prayer, a heartfelt expression of resignation made all the more touching by its massed choruses and the gently churning pipe organ accompaniment that propels it along. "Marcella" ("Carl and the Passions") and "Funky Pretty" ("Holland") are both noteworthy for their codas—repetitive sequences fleshed out into little vocal/instrumental clockworks that are both static and remarkably variegated (imagine a combination of Vivaldi, the Coasters and Steve Reich and you'll have the general idea).

Wilson participated very little in the making of most of these records and when "15 Big Ones," with more of his songs than any album in years, turned out such a disaster, some despaired for his future. It was a surprise, therefore, when "The Beach Boys Love You" was issued in 1977 with more than a dozen new Brian Wilson songs, most of them pretty good. But *strange*—and childlike. There are songs about roller skating children, honkin' down the gosh-darn highway, picking up a baby and the solar system. The last song, by the way, has some of rock's dumbest lyrics ("If Mars had life on it/ I might find my wife on it") and even the album's best selection, "The Night Was So Young," is marred by the stunningly inelegant phrase "Love was made for her and I." "The Beach Boys Love You" is an album about which there is not much middle ground. Detractors find it infantile, beneath contempt, proof positive of Wilson's final burnout; admirers acknowledge the record's childishness but discern a clumsy freshness, a redeeming sense of wonder about it all. Still, it's a long way down from "Pet Sounds."

During the decade between "The Beach Boys Love You" and Wilson's first solo album (entitled, simply, "Brian Wilson"), the Beach Boys, with and without Brian, became an immensely successful oldies act, but their occasional new releases had little or nothing to recommend them. "She's Got Rhythm" ("The M.I.U. Album") was an addlepated but irresistably catchy dance number about a close encounter in a discothèque. And "Good Timin'," included on the "L.A. Light Album" (1979) is probably the Beach Boys' last great single, but Brian had written it several years earlier; with its lush harmonies, smooth production and close adherence to commercial formula, it could have easily fit onto "All Summer Long."

"Brian Wilson," released in 1988, is a partly satisfying comeback. Wilson seems in shaky but unmistakable form (no fewer than seven doctors are credited in the liner notes). "Love and Mercy," the opening track, was a brilliant single—concise, distinctive, immaculate and not a second too long—and should have been a colossal hit, with "Meet Me in My Dreams Tonight" the natural followup.

But Wilson's disturbances are still evident throughout the album. "A lot of people out there hurtin' and it really scares me," he confesses at one point. He compensates for a loved one by retreating to a fantasy world where "planets are spinning around." In "Rio Grande," the album's final cut, the passage about night blooming jasmine "a'creepin' through my window," scored for lowing chorus and solo voice, fed through a sampler and doubled by harpsichord, is one of the most beautiful and chilling musical evocations of paranoia I know.

Ultimately, Wilson's trajectory must be charted as a journey from innocence to experience and then back into a cocoon of damaged innocence, changed but hardly richened by the voyage out. It is perhaps unfair to call his career a tragedy: He has already given us a body of work that, with its best moments, set new standards for invention and lyricism in a musical genre that was just discovering it was an art form. But when I compare "Pet Sounds" and the "Smile" tracks with what came later, I experience many of the same emotions I felt that day in Hawthorne—awe, desolation and genuine regret that things turned out the way they did.

Wigwag (1990)

6 LISBON TRAVIATAS AND OPERA OBSESSIVES

Renata Scotto, taking the chance of her career, walks nervously onto the stage of the Metropolitan Opera House and prepares to sing. Before she can open her mouth, there is commotion at the back of the hall and then, unmistakably, a voice is heard from the standees: "Boooooo! Evviva Callas!" Backstage, after the debacle, a Miss Piggy doll waits in her dressing room, thoughtfully delivered by hand earlier that afternoon. Scotto's sin? She had dared to sing Bellini's *Norma*, a role once associated with Maria Callas, now gone but definitely not forgotten, particularly by a certain breed of fan.

On Stefan Zucker's Saturday night radio program on WKCR-FM, listeners call in from five states to express their dissatisfaction with the current condition of singing. Established sopranos are regularly excoriated as "filth," "sluts" and "pigs." Zucker, who bills himself as the "world's highest tenor," publishes a magazine called *Opera Fanatic*, which consists mainly of his own extreme but knowledgeable reviews of recent performances and lively investigative reporting of a maniacal bent (Is Aprile Millo really the illegitimate daughter of JFK? Why did Ileana Cotrubas walk out of the Met? It's a *National Enquirer* for operamanes, with less than one-thousandth the circulation.)

Although she is not actually a character in the play, Maria Callas permeates the action of Terrence McNally's engaging *The Lisbon Traviata*, currently in production at the Promenade Theater in Manhattan. Mendy, a convinced and flamboyant operaphile, is told of the existence of a pirated recording of Callas singing *La Traviata* in Lisbon. He becomes hysterical—whining, wheedling, teasing and threatening friends and strangers alike—in an effort to get a copy of the recording immediately. "Don't you know that this is the most important thing in the world to me?" he asks desperately.

Welcome to the world of the opera obsessive, a small but (ahem) *vocal* subculture known to every singer, manager, administrator and critic concerned with music theater. What unites them is a passion for opera that is profound and subjective, in which the intense and sometimes tawdry stories of love and betrayal (both onstage and off) ring with personal urgency.

It is a small, tight clique. Opera obsessives meet in the classical department of Tower Records, in the standing-room section of the Met, at selected performances in San Francisco, Philadelphia, Chicago, Seattle, Santa Fe and other venues around the country. Pirate companies issue foggy recordings, taped in a briefcase, of forgotten operas and know that they will find buyers. Retired divas dote on these fans, who know their records, their histories, the anniversaries of their debuts, and pay appropriate (and often *more*-than-appropriate) homage.

The Italian repertory of the nineteenth and early twentieth century is the core literature for opera obsessives. Mozart and Monteverdi are too temperate, too chaste, and Wagner's works do not divide easily into arias. Recent music is beneath contempt, of course. This leaves the operas of Bellini, Donizetti, Rossini, Verdi, Puccini and such lesser lights as Cilea, Mascagni, Leoncavallo, Catalani and the gang. Blood and guts are important to the obsessive, effusive emotionalism crucial.

These are not the erudite aficionados of music theater—the "buffs"—who objectively discuss the merits and demerits of, say, unusual productions and new operas. In general, the future of opera holds little interest for the obsessives; they are much more interested in its past, and the current manifestations of that past. Nor should they be confused with the straightforward, middle-brow music lovers who just want to "hear Pavarotti sing"—something, anything. No, an obsessive's knowledge is deep as a well and, often, just as narrow.

Not long ago, an opera quiz book was published that, along with many legitimate queries, asked the immortal question: "What is the name of Renata Tebaldi's poodle?" (*You don't know?* That can get you ostracized in this circle.) Opera obsessives know who sang the world premiere performance of *Cavalleria Rusticana*, the dates of soprano Claudia Muzio, the Met schedule for the next two years. They exult when yet another Donizetti opera is pulled from an opera-house basement (increasing the tally to nearly seventy, of which some four or five have found a regular place on the stage). They know the ushers at the Met (and, if they are wealthy enough, those at La Scala as well) and like to talk of the opera powers loudly and publicly, in a manner that suggests intimacy.

Met music director James Levine, for example, is always "Jimmy" to the standees. And then there are the nicknames—Joan Sutherland, a good ol' girl from Australia, has been dubbed "La Stupenda" and is spoken of as such with a perfectly straight face. And Callas, mercurial and brilliant dramatic soprano and consort of Aristotle Onassis, has been elevated to the position of secular icon. She is—get ready—"La Divina." One begins to see why dispassionate critical discourse is impossible with an opera obsessive.

As it happens, there *are* some critical works aimed specifically at this audience. One is Ethan Mordden's catty, amusing and literate *Demented: The World of the Opera Diva.* Less successful is the late Lanfranco Rasponi's *The Last Prima Donnas*, which, for the uninitiated, really tests the gag reflex.

Example: "It was with great anticipation that I called on [soprano] Gina Cigna in her stately apartment in Milan: She had sent shivers down my spine during my youth in Florence, where she sang year after year in a variety of roles. I did not expect anyone so animated, down-to-earth and totally honest, full of that particular form of witchcraft called charm. Within a matter of minutes I felt as if I had always known her: Her eyes, which her colleague Elena Nicolai had described as 'fantastic—like stars,' spoke as expressively as her words."

And so forth, more than fifty singers sketched in the gooiest of press-agent prose, all of them simply *wonderful* people, all of them bemoaning the decline of the West (which just happened to correspond to the date of their retirement), most of them haughtily and ignorantly dismissive of contemporary music.

And there's the rub, for anybody who cares about opera as a vital, ongoing art form. Most opera cultists wear their ignorance proudly, a badge of reaction. Recently, during one of the Metropolitan Opera radio quizzes, panelists were asked to name an opera that had saints in it. The "experts," obsessives all, hemmed and hawed ("Gee, wasn't there something by Massenet?") and came up short, completely ignoring the late Virgil Thomson's *Four Saints in Three Acts* (which, despite its title, is cast in four acts and features more than thirty saints) and seemingly knowing nothing whatever of Olivier Messiaen's *St. Francis of Assisi*, a five-hour opera by the most important composer France has produced since Maurice Ravel.

Meanwhile, in *The Lisbon Traviata*, McNally's Mendy contemptuously dismisses Stephen Sondheim, Glass and Bruce Springsteen, all of whom, to this taste, have written much better music than anything by Cilea. Knock the contemporary, knock the popular, knock the American—three rules for the compleat opera obsessive, and for a cultural straitjacket.

But one need not be an obsessive to enjoy *The Lisbon Traviata;* it is an amusing and, in the end, affecting play. Oddly, for all of its arcane references, there are several errors that no self-respecting opera fanatic would ever make. Callas was 53, not 54, when she died. The names "Turandot" and "Don Jose" are consistently mispronounced throughout the play. The record store Music Masters is on West 43rd Street, not 44th Street. Another store mentioned, the lamented Discophile, has been out of business for five years; its phone number, still imprinted on my brain as I'm sure it is on the brains of hundreds of other music lovers, was 473–1902, rather than the one quoted in the play. These are trivial points, but, in a work that is so proudly concerned with minutiae, they ring false.

"We all love, we're all going to die," observes Mendy. "Maria knew this." Well, yes, and so does my second cousin. Still, there is much that is tender and true. "How do you describe a miracle to somebody who wasn't there?" Mendy asks, turning up the music, wonder and rapture in his voice. "Do yourself a favor and put on one of her records." Right, but how about giving Maria a last name this time?

Newsday (1989)

7 A PLEA FOR HESITATION

Earlier this season, I sat with a rapt audience at Carnegie Hall, listening to Mahler's Symphony No. 9. Written during the composer's final illness, the symphony provides one of the most difficult, haunting yet ultimately rewarding musical experiences in the repertory. When the final notes of the closing Adagio dissipate into air, the silence should be as eloquent as the sounds have been—a charged residue, if you like; a silence that is a musical statement in itself, transfigured by what has gone before. At Carnegie, Mahler's silence, so hard won, lasted barely a second. Somebody shouted "Bravo!," the audience responded reflexively, and one left the hall frustrated, rudely shaken from an engrossing dream.

Premature applause is one of the most disturbing elements of our musical life. I have long despaired of hearing the final notes of the first act of *Der Rosenkavalier* in an opera house. It is a wonderfully poignant moment, and it is inevitably interrupted. The Marschallin sits at her mirror, worrying about her lover, feeling mortal. The curtain slowly starts to fall, and the exquisite cadence Strauss created to accompany its drop is suddenly lost in applause. Ironically, the better the performance has been, the more quickly it is likely to be disrupted.

Will I ever attend a performance of Schubert's "Trout" Quintet in which the playful false ending in the last movement isn't broken by ill-timed cheering? The fans vie to shriek the first "Brava!" at diva concerts, and a recent performance of Debussy's "Clair de Lune," was marred by clapping the moment the final note was played—the musical equivalent of a photo finish.

Music is not a sporting event but, at its best, a taste of the sublime; the most abstract of the arts, it is, paradoxically, the most human. Jean Sibelius asked that his Symphony No. 4 be followed by no applause whatsoever, and that the audience leave in thoughtful dignity. Glenn Gould once titled an article "Let's Ban Applause!" Most of us would not go this far, but applause can be ruinous unless it is carefully considered.

Applause between movements or sections of a song cycle is bad enough (I recently attended a rendition of Schumann's "Frauenliebe und Leben" in which *every single song* was bravoed, handily sapping the score's collective power), although a polite note in the program guide or a shake of the head from the artist will usually discourage such transgressions. But what can one spectator do to still premature applause without seeming to shush, hector and hamper one's neighbors, who have, after all, paid for their tickets and are legitimately entitled to express their pleasure?

A plea: Hold the applause until the music has completely died away. Wait for that relaxation of a conductor's shoulders, the drop of a pianist's hands or the easy grin that releases singer from song cycle. And then applaud as long and loudly as you wish. Your enthusiasm will then complement the music, rather than diminish it.

The New York Times (1986)

8 LIVING WITH RECORDINGS

More people have now experienced Beethoven's symphonies, Verdi's operas, and Bach's concertos in their home than will ever set foot in a recital hall. Particularly for those who have grown up in the era of the long-playing album, discs have largely replaced the live concert as the most feasible and economically viable medium to encounter either a composer's work or a performer's re-creation.

This is not necessarily the best of all possible worlds, but it is the one we live in. The recording has altered our perception and we have changed our listening habits. And while it is doubtful that we will want to see many films again and again—at least not in immediate succession—nor instantly reread a book, we have no problem with repeated listenings to favorite recordings. Music is the most abstract, the least narrative of the arts, yet even in the concert hall, we might become annoyed if confronted with the same work three evenings in a row. But somehow with recordings it never matters. For whatever reason we listen—the music, the performance, the sonic splendor of today's advanced recordings—we can replay certain discs indefinitely.

A performance we choose to live with at home may well be different from one we would applaud in the concert hall. During the New York Philharmonic's Mahler cycle in 1976, James Levine conducted an incandescent Eighth Symphony at Carnegie Hall, and immediately preceding the final earthshaking chords, the conductor interpolated a brief rest. Levine's pause was not in the score, yet it was brilliant theater and added an extra degree of excitement to what was already an impressive performance; it couldn't have been more than a couple of seconds long, yet seemed an eternity. On repeated listenings, however, with the element of surprise missing, I would not want to live with this innovation, and would choose a more orthodox rendition—that of Bernard Haitink, say, or Leonard Bernstein.

Few of us would agree with the critic who suggested that Mahler's next symphony, the Ninth, be boiled down to the twenty minutes he thought it deserved, nor with Richard Strauss, who argued that the early Verdi operas should be dismembered and served in a potpourri on German opera stages. But there is still the chance that—in the privacy of our own home—we will do exactly what we profess to despise. Though we may disdain those ubiquitous ads for "Fifty Great Masterworks of Music" offered on late-night television, promising Haydn's 93rd Symphony in as many seconds, or Chopin's "Minute Waltz" in double time, nevertheless, we tend to compile our *own* "Top 40" list no matter how large our record collection. Time and a certain aural hedonism entice us to abandon integral sweep and chronological structure—we want to hear our favorite parts, and in a hurry.

For instance, on a recent evening, I played Side Eight of *Der Rosenkavalier,* the opening chorus from Bach's Sixth Cantata, all of Franz Schmidt's "Book of the Seven Seals" and the last half of the last movement of Busoni's Piano Concerto—twice. We have learned to experience Wagner's "Ring" cycle twenty minutes at a time, divorced from its dramatic and philosophical overtones, and while it may take me a moment to remember that the great love duet from *Les Troyens* occurs in Act Four of Hector Berlioz's opera, I instantly recall that it takes up most of the eighth side of the Philips recording.

Those with the financial resources to consider opera primarily a theatrical experience will consider this sacrilege. But though I have listened to *Les Troyens* several times—five records at a single sitting, armed with score, *Aeneid* and Jacques Barzun's biography of the composer—and would welcome the opportunity to experience the pageantry and spectacle of a complete staged performance, the fact remains that necessity dictates that I generally take my Berlioz a side at a time.

What are some records that will wear well at home over a long period of time? Everyone will have their own nominations. Performance and sound quality have much to do with our choices, yet some of our favorite recordings may be found wanting. For example, there are a number of contemporary recordings of Jean Sibelius's Sixth Symphony, but none of them remotely captures the sunny radiance of this vastly underrated work as does George Schneevoigt's version from the early thirties, now reissued on Vox Turnabout (Turnabout THS 65067). This serene, elevated work, for me the most affecting music Sibelius wrote, is captured beautifully by Schneevoigt, in a performance of warmth, understatement and flow. Yet the sonic quality is not of the best, and the Finnish National Orchestra performs sloppily at times. But we become used to poor sound, and flubs are forgivable. In a time before tape splicing, note perfection was difficult to achieve. An entire generation, weaned on recordings, grew up expecting every horn player to make a mess of his part in Beethoven's Ninth Symphony as does the soloist in the early Weingartner performance, and, for many, Alfred Cortot's many finger slips are an integral part of early memories of Chopin's music.

The obvious deficiencies here have never kept us from enjoying these great renditions; one might as well complain that the *Canterbury Tales* are incomplete, or that the Venus De Milo has no arms.

This an age of musical pluralism, and there is much new material being composed in a variety of different styles. As a result, we often have to learn a composer's language before we can fully appreciate what is being said. Nobody benefits from repeated rehearings more than the so-called "minimalist" composers. For the music written in this genre, on first hearing, may strike some listeners as sheer repetition, along the lines of a broken record. Oddly enough, the best way to correct this mistaken impression is yet another hearing; on the second listening, the music will already sound less simple, and one will begin to appreciate the fertile activity beneath the repetitive surface sheen. Only after several hearings can the charm, subtlety and formal structure of a work such as Steve Reich's "Music for Mallet Instruments, Voices and Organ" begin to reveal itself. The composer's own recording of this hauntingly beautiful work is definitive (in a three-disc set DG 2740 106, including other works), and I find myself able to sit through several plays with no flagging of interest. Although quite enough has already been said about the Eastern influence on minimalist music, one is still tempted to use the analogy of a mantra—the more one concentrates on what initially seems unchanging, the more variety one is aware of. The impeccably restrained rhythmic climax to "Mallet Instruments" is one of the most quietly enthralling moments on record.

Thorny works which cannot possibly be grasped at a single hearing have found their best friend in the recording medium. Bach's "Goldberg Variations"—created for the insomniac Count Kayserling, who requested some music to carry him safely through the night—represent the composer at his most concentrated and impenetrable, and do not easily give up their secrets. The popularity of the "Goldbergs" is largely a recent development, and I think it can be traced to our ability—through discs—to spend the time and energy necessary to find our way to the heart of this extraordinary work. Glenn Gould's 1955 rendition of the "Goldberg Variations" (Columbia 31820) is a performance of intelligence, originality and fire. I continue to find fresh delights on every rehearing. Gould's recording of the "Goldbergs" has a strange chemistry to it, which seems to change valence depending on the time of day one listens to it. If I put on Gould's disc in the morning, I am invigorated by the variations in quick tempo, by their propulsive energy and elemental verve. Heard after midnight, however, Gould's disc is a different experience altogether. The vivacity of the brisker movements remains attractive, but I find greater interest in the somber, mysterious slower passages. In the extraordinary 25th Variation, the listener senses the ghostly presence of every insomniac back to Kayserling. Nobody has ever captured the vulnerable hush, the whispery despair of a sleepless blue night becoming haggard gray morning as does Gould in this recording. It is best heard during the hours for which it was created. Again and again, and as we choose.

The New York Times (1982)

9 A RETURN TO TANGLEWOOD

Lenox, Mass.

Every town has secrets known only to its young people—the hideout clearing in the pines, the abandoned storm cellar off the main road, exactly which stones in the wall can support a sprinting 10-year-old—passed on like folklore from older children to the next generation.

If we revisit a place often enough, every return comes to seem nothing less than a summation of our massed experience there. "Time present and time past are both perhaps present in time future, and time future contained in time past," T. S. Eliot wrote in "Burnt Norton"; stepping onto the grounds of Tanglewood, I am overcome by my own history—time past, present and future.

1970. I am fifteen years old, child and man, away from home for the first time, staying up till dawn, discussing everyone from Frederic Chopin to Frank Zappa with fellow students at the Tanglewood Music Center in Lenox. The soft beauty of the ancient mountains, the sustained, intimate contact with artists from around the world, the consecrated professionalism that pervades one's experience there—all help instill the sense that music can be something on which it is possible to build a life. Suddenly I have peers who understand (and sometimes share) my obsessions, with whom I can talk of the pieces I am learning on the piano, the compositions I am trying to write, obscure recordings, the proper way to dot a sixteenth note and the dream of what Glenn Gould called "the purpose of art. . . . a gradual, lifelong construction of a state of wonder and serenity."

There is Aaron Copland in the library, elucidating his "Piano Variations" to adoring acolytes. Leonard Bernstein, tanned and vigorous in an open-necked pink shirt, not yet the grand old man he later became but already legendary, walks through the main gate, and cafeteria talk comes to a temporary halt. He waves in our general direction, and such is the Bernstein charisma that everyone at our table feels specifically singled out by his greeting.

Now, twenty-one years later, I have returned to Tanglewood for a week—to wander the grounds, observe the classes, meet the students, renew acquaintance with some of the professors, take in as much as I can. A great deal will necessarily be missed—master classes conflict with chamber music performances, which may themselves preclude a sectional rehearsal of the student orchestra. But, on the last day of June, more than one hundred young musicians associated with the Tanglewood Music Center and the Boston University Tanglewood Institute meet in the Theater-Concert Hall for the opening exercises; to sing once again the Randall Thompson "Alleluia" (composed for the first of these occasions fifty-one years ago); to listen to an array of speakers call up the spirits of summers past.

Serge Koussevitzky, music director of the Boston Symphony Orchestra from 1923 until his death in 1951, founded what was then the Berkshire Music Center at Tanglewood in 1940. He is buried just outside the center of Lenox; his home, Serenak (one of those vast, elevated mansions improbably referred to as "Berkshire cottages") has been purchased and lovingly restored by the Boston Symphony. At the ceremony (which is *very* Boston: traditional, academic, proudly insular, complete with a quote from Ralph Waldo Emerson—J.P. Marquand would have loved it), Koussevitzky's legacy is invoked by speaker after speaker: Seiji Ozawa, the BSO's current director; Phyllis Curtin, a distinguished soprano, teacher and dean of Boston University's School of the Arts; and Leon Fleisher, a pianist and conductor who now serves as the artistic director of the TMC.

But "Koussy" (as he is called even by those born two decades after his death) is not the only spectral presence in attendance this year. Copland, who long served as dean of the faculty, died last December. And Bernstein, who died in October, was one of the center's first students and grew to become its most celebrated alumnus. He steadfastly maintained his connection with Tanglewood over the course of half a century, returning every summer to instruct and inspire the young musicians. Here, on Aug. 19, 1990, he conducted his last concert—a Boston Symphony program featuring selections from Benjamin Britten's *Peter Grimes* and an extraordinary Beethoven Seventh Symphony.

"The ghost for us is not Koussy," Mark Stringer, a twenty-seven-year-old fellow in the conducting program, tells me as we sit in the Tanglewood Shed following a TMC orchestra rehearsal. "The ghost for the younger generation is Lenny."

Stringer, soft-spoken, intense and articulate, served as Bernstein's assistant conductor from 1988 to 1990; this is his second summer at Tanglewood. When the mortally ill maestro was too weak to supervise some scheduled rehearsals, Stringer was tapped to prepare the TMC orchestra in Copland's Third Symphony. "He was terribly supportive of his students' early progress," Stringer says. "Just like a parent, you know—what a *nice* crayon drawing you made for me! Later on, if he thought you could handle it, he'd tell you ways that your performance might have been improved. But, at the time, he just gave you a kick in the pants to get you out there and then wrapped you in his arms when it was all over and you were safely backstage again. You felt for a moment that you were the only person in the world who mattered to him. He took such pride when one of his students really *understood* something."

Brynn Albanese, twenty-two, who served as the concertmaster for the initial TMC orchestra concert, and Sophie Willer, twenty-one, a cellist from Vancouver, Canada, missed the opportunity to work with Bernstein. "One summer too late!" Albanese says with a mixture of regret and exasperation over a luncheon in Lenox. But both were dazzled to be in close contact with Ozawa, Fleisher and the other conductors and teachers who will pass through the Berkshires this summer.

Both women have determined to make careers in music. Willer simply says that she wants "everything—chamber music, concertos, the solo repertory, to play in a great orchestra." Albanese is more specific: "I want to be a concertmaster, the first woman concertmaster—maybe even the first woman player—that the orchestra has ever had. And then I'll show them all." (Despite overwhelmingly male bastions such as the Berlin and Vienna Philharmonics, many leading symphonies are beginning to achieve some parity of representation and, this summer, all but three of the 27 violinists in the TMC orchestra are women.)

"I've never played in a better orchestra in my life," Albanese says. "We're better than a lot of famous groups I could name—although I probably shouldn't." (She proceeds to name two and immediately puts her comment off the record.)

"The TMC orchestra is a professional-level ensemble," she continues. "Ozawa just assumes that we're all pros, and then he really works us."

"You don't even have time for solo practice," Willer adds. "I'm glad I did a lot of that in the weeks before I came up, to get into shape. Here, you're burned out all the time—what with all the rehearsals, the classes, the concerts—but it's a thrill to be burned out in this way."

Robert Olivia, a twenty-seven-year-old clarinetist based in Boston, has worked out a practice arrangement with his roommate, a German cellist. "Basically, we're considerate of one another. His instrument is a lot heavier than mine, and I don't like seeing him have to lug it around. So I look for another place to play when he's in the room."

There is still time to socialize. Young musicians work hard, but often play harder. Five Chairs, on the Lenox-Pittsfield Road, is the bar of choice this year, and the dorms are full of extramusical intrigue. "We've got some real romantic soap operas going on already," Albanese says. "And we've had some great parties." At a recent TMC fete, Albanese acted as disk jockey. What did she play? "Oh, the Gipsy Kings. African bongo drum music. Madonna. You know." (An upcoming post-concert party with "Mr. Ozawa"—my generation, behind his back, inevitably called him "Seiji"—will, she promises, be "much more sedate.")

"Sedate" was hardly the appropriate word for an Independence Day celebration featuring Bob Dylan—the first concert in the Tanglewood Shed this season. All day long, the caravans streamed into Lenox, effectively immobilizing the town. The concert itself was a loud, raucous, rock and roll *event*, with all the excitements and irritations that generally accompany such festivities. By the third song, the stage had been rushed—the determinedly unthreatening volunteer usher simply threw up his hands and got out of the way—and it was necessary to stand on one's chair to get even a glimpse of the Star Attraction. ("Survival of the tallest," a vertically challenged friend muttered to herself.)

Yet the chaos that ensued was basically a gentle one. People queried politely whether their pot smoke would be intrusive (without, it must be admitted, really wanting an answer) and apologized abjectly when they stomped rhythmically on your foot. Dylan himself, however, was, (to employ a useful, antiquated Americanism) a "bummer." Backed up by a generic band that could have been playing behind anybody, anywhere, he raced—all but *rapped*—through a set of songs that were practically indistinguishable from one another, with unfamiliar arrangements, unfamiliar emphases and, more than occasionally, unfamiliar melodies. And so the audience joined forces to play a collective round of "Name That Tune." ("I heard something about a brass bed—must be 'Lay Lady Lay.' " "That impossibly false, silly choice he's offering us—w-why, it's 'Gotta Serve Somebody.' ")

The lawn is reportedly a shambles at the end of the concert, but all is immaculate by ten the next morning; the crew has obviously been working overtime. Out at the main gate, the ticket windows are already busy; the Shed is pretty much sold out, but it is still possible to obtain admission to the grounds. On wet nights, the Tanglewood lawn becomes a soggy sea of umbrellas, but if it is fair (and the visitor is willing to put up with what must be a greater density of mosquitos per cubic foot than any place north of the Amazon), one may commune with the trees, the stars, the cool night air (I have known the temperature to fall 30 degrees in only a few hours) accompanied by some beloved music that seems to literally emanate from the surroundings.

Throughout the day, as the visitors from New York, Boston and points in between find their way to Lenox, classes continue in a dozen places on the Tanglewood grounds. Phyllis Curtin's master class in voice is justly famous. It is not unusual for her to spend the better part of an hour on a single phrase—polishing it until it shines, until all of its levels of meaning are uncovered, until words and music are united in organic symbiosis.

Albanese is occupied throughout the morning with the Brahms A-major sonata for violin and piano, under the direction of Julius Levine. In a nearby studio, Louis Krasner, now in his eighties and the man directly responsible for the creation and first performances of two great violin concertos (by Alban Berg and Arnold Schoenberg), is working with five musicians on a performance of Mozart's String Quintet (K. 515). Contemporary music is not slighted: During the first week, works by John Corigliano, John Cage and Elliott Carter are in preparation for upcoming concerts.

Seiji Ozawa is in perpetual motion throughout the day. In the morning, he rehearses with the Boston Symphony for three hours, readying the Brahms "German Requiem" for Sunday afternoon. At 1:30 he works with the TMC orchestra. He does not patronize the students; he demands as much from them as he would from the BSO and is rewarded with an additional surge of enthusiasm, the natural byproduct of youth and untested passion. At 9 he conducts the BSO in Gustav Mahler's "Resurrection" Symphony—an evening-length work by the composer with whom Bernstein felt the most profound spiritual affinity. Indeed, it is impossible to imagine the great Mahler revival of the sixties proceeding as it did without Bernstein's advocacy; and the entire first set of symphony concerts has been consecrated to his memory.

The following night, the TMC gets its hour to shine, and Albanese's contention is borne out: Formed only one week before, this is already a brilliant orchestra.

Afterward, there is a party over at Highwood, an adjoining mansion that now houses the Tanglewood press office and various other administrative facilities. The mood among the students is one of jubilation—"We did it! We really did it!"—and if the party begins sedately, its continuation (at Five Chairs and back in the dorm) becomes lively indeed.

"There are two sorts of students here," clarinetist Olivia says. "Most of us are wide-eyed and exhilarated. Others—mostly those who have been here before and gotten used to seeing...oh, Seiji Ozawa walking across the lawn—are more blasé. If I come back next year, I hope I can retain the spirit of excitement I feel today."

I've come back—as student, observer and, finally, critic—almost every summer since 1970, and the excitement has never left me, although it is now tinged with nostalgia and the memories of many years.

On my last day at Tanglewood, before the afternoon concert, I decide to pay an impromptu visit to Wheatleigh, a Berkshire cottage even grander than Serenak. At the turn of the century, the estate was built by the Countess de Heredia, another in a long line of Lenox "characters." In the last two decades, it has been an arts center, a private house, a Tanglewood residence, a mildewed basement bar called La Cave and now a luxurious resort and restaurant; I have known it in all capacities. Ten years ago this month, I shared a bottle of wine on the sloping back lawn of Wheatleigh with a friend; we felt young and enchanted, kicked off our shoes, toasted the summer's crescendo and generally behaved as if we'd fallen into one of *The New Yorker*'s toniest advertisements. Wheatleigh encourages such fantasies.

No Wheatleigh administration has yet disturbed the grotesquely charming corner where a dozen or so of de Heredia's small dogs lie moldering under mossy, weatherbeaten tombstones. For years, the site was famous locally as the "Poodle Tower" (in fact, the dogs were apparently Pomeranians, and the tower burned last year). But Bo-Peep, Missy and the other canine mummies remain—along with a tablet commemorating the unfortunate pup who had the bad taste to die in Nice.

Another secret to which only a young person would be privy; nobody takes visiting music critics to the "Poodle Tower." Wandering up the Stockbridge Road, I can hear the Boston Symphony beginning to tune, and I realize that in another twenty-one years—which will pass even more quickly than the time since I first came to Lenox, gangly and awed—it will be 2012; I shall be fifty-seven years old. The recognition is unsettling, but the day is fine, the summer has just begun, and I enter the Tanglewood Lion Gate both happy and aware that I am happy, eager for secrets yet to be divulged.

Newsday (1991)

ACKNOWLEDGMENTS

I've written four or five lengthy drafts of this general acknowledgment and finally given up on it. More than any of my books, this feels like the summation of a lifetime (to the age of thirty-seven, anyway) and I want to get everybody in. But it just won't work, not without adding two or three more pages of autobiography to what is already a pretty personal book. So I have cut the list dramatically, and most of the people who should by all rights be thanked won't be, this time around. You know who you are, and you have my gratitude.

Still, I must acknowledge friends and editors at the *Soho News, The New York Times* and *Newsday*—especially June Carlson, Lois Draegin, Josh Friedman, Arthur Gelb, Gerald Gold, Sylviane Gold, Peter Goodman, Annette Grant, William H. Honan, Allen Hughes, John Leese, Tony Marro, Caroline Miller, Peter Occhiogrosso, Noel Rubinton, Howard Schneider, Harold C. Schonberg, Marvin Siegel, Phyllis Singer and Thomas J. Wallace. Thanks also to Sheldon Meyer, my editor at Oxford University Press; Arnold Goodman and Elise Simon Goodman, my agents; Hilary Dyson, Jeffrey Herman and Allan Kozinn for their help in the preparation of the manuscript; and John Rockwell, the most intrepid and original music critic of his generation, who served first as inspiration, and later became mentor, teacher, colleague and friend.

I owe a profound debt to Leonard Altman, without whose empathic kindness and endless patience, this book—and, quite possibly, its author—would not exist.

Thanks to my father and mother, Ellis and Elizabeth Page, for reading through these articles and providing insightful, diplomatic commentary. Thanks also to my son William Dean Page, now five years old, for being so goodhearted when I guided his fingers from the computer terminal, and to infant Robert Leonard Page, for his long naps and cheerful awakenings.

For all of my work, I am indebted to the University of Connecticut Music Library, and particularly to Gloria A. Sterry and Dorothy

Bognar, the chief librarians there during my childhood. Their willingness to bend the rules permitted a youngster access to the music he loved; had it not been for this accommodation, I should likely be in another field. And I am grateful to Charles Jones, my principal composition teacher, for training me to think critically about music—not that he will agree with every opinion in this book!

Finally, thanks to my wife, Vanessa Weeks Page, for everything. Dedicating this book to her is redundant: She has helped shape every paragraph.

The following articles first appeared in *The New York Times* (some of them under different titles): "Geraldine Farrar," "Steve Reich," "Ellen Taaffe Zwilich," "A Plea for Hesitation," "Living with Recordings," "A Suzuki Orchestra," "'Postcard' from Juilliard," "A Visit to California," "Robert Taub Plays Babbitt," "Pavarotti at the Garden," "'Bleecker Street' in Charleston," "Bach and the Rodents," "John Cage at Barnard," "Piano and Electronics: Rebecca La Brecque," "The Return of Conlon Nancarrow," and portions of "Leonard Slatkin."

The following articles first appeared in *Newsday* (some of them under different titles) and are reprinted by permission: "Marin Alsop," "Leonard Bernstein at 70," "Leonard Bernstein: In Memoriam," "Mieczyslaw Horszowski," "Van Cliburn," "Christoph von Dohnanyi," "Nadja Salerno-Sonnenberg," "Maria Bachmann," "Matthew Epstein," "Christopher Keene," "Maurizio Pollini," "Prodigy: The Enduring Mystery of the Musical *Wunderkind*," "Meredith Monk at BAM," "*Sweeney Todd* at City Opera," "Harry Partch in Philadelphia," "'Swing' and Swados" "Irving Berlin at 100," "Ellington's Orchestral Music," "Arvo Pärt and his 'Passio,'" "Pogorelich at Carnegie Hall," "Ernst Krenek and John Coriligiano," "Who Is Schubert," "Bernstein Conducts Copland," "New Music America at Ten," "*Holy Blood and Crescent Moon*," "Vladimir Feltsman," "The Kronos Quartet," "Peter Serkin Plays Commissions," "A Moment of Silence," "Peter Sellars' *Figaro*," "Mitsuko Uchida's Mozart," "*The Death of Klinghoffer*," "Opening Night at the Met," "The Mozart Bicentennial," "Lisbon *Traviatas* and Opera Obsessives," "The New Age of Brian Eno," "George Antheil Revived," "A Return to Tanglewood," and portions of "Philip Glass" and "Leonard Slatkin."

Portions of "Glenn Gould" first appeared in *Vanity Fair*, while my conversation with the pianist was initially published in the *Soho News* and later expanded for the *Piano Quarterly*. "Anton Webern" was first published in the *Soho News*.

Portions of "Philip Glass" first appeared in *Opera News*.

"Early Music, Lately" first appeared in *Elle*.

"Henryk Mikolaj Górecki" and "Brian Wilson" first appeared in *Wigwag*.

"A Conversation with Philip Glass and Steve Reich" first appeared in *Cover*.

"John Cage" and "Milton Babbitt" were first published, in different form, in *Boulevard*.

INDEX OF NAMES